T0093400

PHYTOCHEMISTRY OF PLANTS OF GENUS *CASSIA*

Phytochemical Investigations of Medicinal Plants

Series Editor: Brijesh Kumar

Phytochemistry of Plants of Genus *Phyllanthus*
Brijesh Kumar, Sunil Kumar and K. P. Madhusudanan

Phytochemistry of Plants of Genus *Ocimum*
Brijesh Kumar, Vikas Bajpai, Surabhi Tiwari and Renu Pandey

Phytochemistry of Plants of Genus *Piper*
Brijesh Kumar, Surabhi Tiwari, Vikas Bajpai and Bikarma Singh

Phytochemistry of *Tinospora cordifolia*
Brijesh Kumar, Vikas Bajpai and Nikhil Kumar

Phytochemistry of Plants of Genus *Rauvolfia*
Brijesh Kumar, Sunil Kumar, Vikas Bajpai and K. P. Madhusudanan

Phytochemistry of *Piper betle* Landacres
Vikas Bajpai, Nikhil Kumar and Brijesh Kumar

Phytochemical Investigations of Genus *Terminalia*
Brijesh Kumar, Awantika Singh, and K. P. Madhusudanan

Phytochemistry of plants of Genus *Cassia*
Brijesh Kumar, Vikas Bajpai, Vikaskumar Gond, Subhashis Pal, and Naibedya Chattopadhyay

For more information about this series, please visit: https://www.crcpress.com/Phytochemical-Investigations-of-Medicinal-Plants/book-series/PHYTO

PHYTOCHEMISTRY OF PLANTS OF GENUS *CASSIA*

Brijesh Kumar
CSIR-Central Drug Research Institute (CDRI), Lucknow, India

Vikas Bajpai
CSIR-Central Drug Research Institute (CDRI), Lucknow, India

Vikaskumar Gond
CSIR-Central Drug Research Institute (CDRI), Lucknow, India

Subhashis Pal
Emory University, Atlanta, Georgia, USA

Naibedya Chattopadhyay
CSIR-Central Drug Research Institute (CDRI), Lucknow, India

CRC Press is an imprint of the
Taylor & Francis Group, an **informa** business

First edition published 2022
by CRC Press
6000 Broken Sound Parkway NW, Suite 300, Boca Raton, FL 33487-2742

and by CRC Press
2 Park Square, Milton Park, Abingdon, Oxon, OX14 4RN

Library of Congress Cataloging-in-Publication Data
Names: Kumar, Brijesh (Phytochemist), author.
Title: Phytochemistry of plants of genus cassia / Brijesh Kumar, CSIR-Central Drug Research Institute (CDRI), Lucknow, India, Vikas Bajpai, CSIR-Central Drug Research Institute (CDRI), Lucknow, India, Vikaskumar Gond, CSIR-Central Drug Research Institute (CDRI), Lucknow, India, Subhashis Pal, Emory University, USA, Naibedya Chattopadhyay, CSIR-Central Drug Research Institute (CDRI), Lucknow, India.
Description: First edition. | Boca Raton : CRC Press, 2021. |
Series: Phytochemical investigations of medicinal plants |
Includes bibliographical references and index.
Identifiers: LCCN 2021019442 (print) | LCCN 2021019443 (ebook) | ISBN 9781032030210 (hardback) | ISBN 9781032030227 (paperback) | ISBN 9781003186281 (ebook)
Subjects: LCSH: Cassia (Genus)--Analysis. |
Cassia (Genus)--Therapeutic use. | Medicinal plants.
Classification: LCC QK495.C1153 K86 2021 (print) | LCC QK495.C1153 (ebook) |
DDC 633.8/34--dc23
LC record available at https://lccn.loc.gov/2021019442
LC ebook record available at https://lccn.loc.gov/2021019443

ISBN: 978-1-032-03021-0 (hbk)
ISBN: 978-1-032-03022-7 (pbk)
ISBN: 978-1-003-18628-1 (ebk)

DOI: 10.1201/9781003186281

Typeset in Times
by MPS Limited, Dehradun

Table of Contents

Preface

Plants are considered to be one of the most important sources of modern medicine. They have been used globally for a variety of ailments since the beginning of civilization. Bioactive secondary metabolites have been considered as a fundamental source of medicine for the treatment of a range of diseases in modern medical system. India has a rich heritage of medicinal plants, Ayurveda, Siddha, Unani, Homoeopathy and Naturopathy. In India, about 6000–7000 plant species are utilized in traditional, folk and herbal medicine. Herbal medicines/ formulations are prepared using crude or processed plants having several active constituents. Knowledge of the phytochemical composition of crude drugs is a very critical aspect in the preparation, safety and efficacy of the herbal product. Identification and determination of the active constituent are now important and crucial prerequisites for the development of modern evidence-based phytomedicine. The use of medicinal herbs or herbal drugs is increasing throughout the world, though one of the main encumbrances in its global acceptance is the lack of quality control or standardization for the establishment of consistent pharmacological activity. A reliable chemical profile or simply a quality control program for the production of herbal drugs can serve the purpose.

Cassia is an indigenous plant in Africa, Latin America, Northern Australia and Southeast Asia. Several *Cassia* species are of highly commercial and medicinal significance since they are used as spices and in traditional medicines. Some of the most widely recognized species of this genus are *Cassia auriculata, Cassia fistula* and *Cassia occidentalis* followed by *Cassia siamea* and *Cassia uniflora. Cassia* species are reported to have various pharmacological activities such as antibacterial, analgesic, anti-inflammatory and antiarthritic, hepatoprotective, antitumor, antifertility, antifungal, antioxidant, antileishmaniatic, antimicrobial, CNS and hypoglycaemic activity. Different class of compounds reported from *Cassia* species are anthraquinones, phenolics, flavonoids, chromenes, terpenes, proanthocyanidins, coumarins, chromones and lignans. The taxonomy and nomenclature of *Cassia* species are quite complex. It is very difficult to differentiate them due to their overlapping morphological characters and close similarities, which usually leads to misidentification and misinterpretation of the components. Genus *Cassia* is in great demand due to its medicinal properties.

Recently, mass spectrometry has played a central role in the field of plant metabolomics. It facilitates efficient analysis of metabolites in the complex matrix of plant extracts due to high sensitivity, selectivity and versatility. LC coupled with tandem mass spectrometry detection (LC-MS/MS) is a sensitive, selective and efficient technique for the comprehensive qualitative and quantitative analysis of plant metabolites. The work contained in this book includes development of a qualitative HPLC-QTOF-MS method to identify 24 phytochemicals in CO extract and validation of a rapid and sensitive UPLC-QqQ_{LIT}-MS/MS method for simultaneous determination of eighteen bioactive compounds in different plant parts of five *Cassia* species viz., *C. auriculata, C. fistula, C. occidentalis, C. siamea* and *C. uniflora*. Principal component analysis (PCA) was applied to study chemical variations of bioactive compounds among *Cassia* species. Moreover, *in-vitro* anti-proliferative activity of these *Cassia* species was also evaluated against five human cancer cell lines viz., human cancer cell lines A549 (Lung carcinoma), MCF-7 (breast adenocarcinoma), DU145 (prostate carcinoma), DLD-1 (colorectal adenocarcinoma) and FaDu (squamous cell carcinoma of pharynx) using SRB assay.

C. occidentalis (CO) Linn (also known as *Senna occidentalis*) is a weed common in South Asia. Leaf and stem of CO are used for treating bone- and joint-related diseases by Indian traditional healers. Such use has now been replicated in a few relevant preclinical disease models. CO contains several compounds including apigenin, isovitexin, kaempferol, quercetin, emodin, 4′,7-dihydroxyflavone, 3′,4′,7-trihydroxyflavone, caryophylline and luteolin that are known to regulate the functions of bone and joint by osteogenic, chondrogenic, anti-adipogenic, anti-osteoclastogenic and immunomodulatory mechanisms. Although, the phytochemical profiles between leaf and stem do not vary much, yet stem extract showed better efficacy over the leaf extract in healing fracture in the rat. Compounds with osteogenic activity in the stem extract were apigenin, isovitexin, emodin, 4′,7-dihydroxyflavone, 3′,4′,7-trihydroxyflavone and luteolin, and several of these have oral bioavailability in the rat. A fraction derived from the total stem extract showed enhanced potency over the total extract due to the enrichment of osteogenic compounds. The osteogenic potency of the fraction was further augmented by a nano-emulsified formulation consequent to increased bioavailability of the osteogenic compounds achieved by the formulation. In preclinical studies, the hydroalcoholic extract of CO leaf and stem was found to be safe. The skeletal effects of CO and its phytochemicals at the cellular, molecular and organismal levels and the potential of CO extract in the treatment of skeletal diseases that lack effective therapy in mainstream medical practice are also discussed.

Acknowledgements

The completion of this book is due to the Almighty who blessed us with all the resources required to accomplish this journey. We are glad to have this chance to express our gratitude to people who have been supportive to us at every time. We are also thankful to Dr. Preeti Chandra for her constant support. We express our deep sense of gratitude to the Director, CSIR-Central Drug Research Institute, Lucknow for his support and Sophisticated Analytical Instrument Facility (SAIF) Division, CSIR-CDRI, India where all the data was generated. Authors thank CSIR-Phytopharmaceutical Mission Project (HCP-0019.1), New Delhi for supporting grant.

Authors

Dr. Brijesh Kumar is a retired Professor (AcSIR) and Chief Scientist of the Sophisticated Analytical Instrument Facility (SAIF) Division, CSIR-Central Drug Research Institute (CDRI), Lucknow, India. He completed his M.Sc. from Lucknow University and PhD from CSIR-CDRI, Lucknow (Dr. Ram. Manohar. Lohia, Avadh University, Faizabad, UP, India). He has 7 books, 11 book chapters and 147 papers in reputed international journals to his credit. His area of research includes applications of the mass spectrometry for qualitative and quantitative analysis of molecules for quality control and authentication/standardization of medicinal plants/parts and their herbal formulations.

Dr. Vikas Bajpai completed his PhD from the Academy of Scientific and Innovative Research (AcSIR), New Delhi, India and carried his research work under the supervision of Dr. Brijesh Kumar at CSIR-Central Drug Research Institute Lucknow. His research includes development and validation of LC-MS/MS methods for qualitative and quantitative analysis of small molecules of Indian medicinal plants.

Mr. Vikaskumar Gond is a research fellow who worked under the supervision of Dr. Brijesh Kumar in the mass spectrometry laboratory, sophisticated Analytical Instrument Facility at CSIR-Central Drug Research Institute, Lucknow. He completed his masters in pharmaceutical chemistry from University of Lucknow. His research includes fingerprinting and characterization of phytopharmaceuticals using LC-MS/MS methods for qualitative and quantitative screening.

Dr. Subhashis Pal is a biologist, born in 1987 in Kolkata, India. He earned his doctoral degree in 2019 specializing in endocrinology and bone biology from Jawaharlal Nehru University, Delhi, India. During his PhD tenure, he worked in the Central Drug Research Institute, Lucknow, India on the potential bone anabolic role of phosphodiesterase inhibitors. Dr. Pal joined Emory University, Atlanta, USA as a staff research associate and currently works at Emory University on the effect of gut microbiome on bone health.

Dr. Naibedya Chattopadhyay is a Chief Scientist at CSIR-Central Drug Research Institute. His areas of research are endocrine pharmacology and metabolic bone diseases. Dr. Chattopadhyay has published >200 research articles and 9 patents out of which 5 are licensed to industries.

List of Abbreviations and Units

°C	Degree celsius
µg	Microgram
µL	Microliter
APCI	Atmospheric pressure chemical ionization
API	Atmospheric pressure ionization
BPC	Base peak chromatogram
CAD	Collision activated dissociation
C. auriculata	*Cassia auriculata*
C. fistula	*Cassia fistula*
C. occidentalis	*Cassia occidentalis*
C. siamea	*Cassia siamea*
C. uniflora	*Cassia uniflora*
CE	Capillary electrophoresis
CE	Collision energy
CID	Collision induced dissociation
CXP	Cell exit potential
Da	Dalton
DAD	Diode array detection
DP	Declustering potential
EP	Entrance potential
ESI	Electrospray ionization
FDA	Food and drug administration
FIA	Flow injection analysis
g	Gram
GC-MS	Gas chromatography-mass spectrometry
GS1	Nebulizer gas
GS2	Heater gas
h	Hour
HPLC	High-performance liquid chromatography
HPLC-QTOF-MS	High-performance liquid chromatography quadrupole time of flight mass spectrometry
ICH	International conference on harmonization
IS	Internal standard
IT	Ion trap
kPa	Kilopascal
L	Litre
LC	Liquid chromatography
LOD	Limit of detection

LOQ	Limit of quantification
LTQ	Linear trap quadrupole
m/z	Mass to charge ratio
mg	Milligram
min	Minute
mL	Millilitre
mM	Millimolar
MRM	Multiple reaction monitoring
MS	Mass spectrometry
ms	Millisecond
MS/MS	Tandem mass spectrometry
ng	Nanogram
NMR	Nuclear magnetic resonance
PCA	Principal component analysis
PDA	Photodiode array detection
psi	Pressure per square inch
QqQ_{LIT}	Triple-quadrupole linear ion trap
QTOF	Quadrupole time of flight
r^2	Correlation coefficient
RDA	Retro-Diels-Alder
RSD	Relative standard deviation
S/N	Signal-to-noise ratio
SD	Standard deviation
t_R	Retention time
UHPLC	Ultra high-performance liquid chromatography
UPLC-QqQ_{LIT}- MS/MS	Ultra high-performance liquid chromatography hybrid linear ion trap triple-quadrupole tandem mass spectrometry
UV	Ultraviolet
WHO	World health organization
XIC/EIC	Extracted ion chromatography

Abstract

Genus *Cassia* is reported to have numerous pharmacological activities such as analgesic, antiarthritic, antibacterial, antifertility, antifungal, anti-inflammatory, antileishmaniatic, antimicrobial, antioxidant, antitumor, CNS stimulant, hypoglycaemic and hepatoprotective activity. A diverse class of compounds reported from genus *Cassia* are anthraquinones, phenolics, flavonoids, chromenes, terpenes, proanthocyanidins, coumarins, chromones and lignans. *Cassia* is a widely distributed indigenous plant found in Africa, Latin America, Northern Australia and Southeast Asia. Several *Cassia* species are of high commercial and medicinal significance and they are used as spices and as traditional medicines. The most widely recognized species of this genus are *C. auriculata, C. fistula* and *C. occidentalis* (CO). *C. occidentalis* Linn (also known as *Senna occidentalis*) is a weed common in South Asia. The leaf and stem of CO are used for treating bone and joint-related diseases by Indian traditional healers. CO contains several compounds including apigenin, isovitexin, kaempferol, quercetin, emodin, 4′7,-dihydroxyflavone, 3′,4′,7-trihydroxyflavone, caryophylline and luteolin which are known to regulate the functions of bone and joint by osteogenic, chondrogenic, anti-adipogenic, anti-osteoclastogenic and immunomodulatory mechanisms. Although, the phytochemical profiles between leaf and stem do not vary much, yet stem extract showed better efficacy over the leaf extract in healing fracture in the rat.

The taxonomy and nomenclature of *Cassia* species are quite complex. It is very difficult to differentiate them due to their overlapping morphological characters and close similarities. This usually leads to misidentification and misinterpretation of the components. New drug formulations from *Cassia* species are regularly being introduced into the market. Knowledge of the phytochemical composition of such crude drugs is a very critical aspect in preparation, safety and efficacy of the herbal product. Now the identification and determination of the active constituents are the crucial prerequisites for the development of modern evidence-based phytomedicine.

Liquid chromatography (LC) coupled with tandem mass spectrometry detection (MS/MS) is a sensitive, selective and efficient technique for the comprehensive qualitative and quantitative analysis of plant metabolites.

The work contained in this book includes development and validation of a rapid and sensitive HPLC-ESI-QTOF-MS method and UPLC-QqQ_{LIT}-MS/MS method for simultaneous determination of bioactive compounds in different plant parts of five *Cassia* species viz., *C. auriculata, C. fistula, C. occidentalis, C. siamea* and *C. uniflora.* Principal component analysis (PCA) was applied to study chemical variations of bioactive compounds among these *Cassia* species. Moreover, *in-vitro* anti-proliferative activity of these *Cassia* species was also evaluated against five human cancer cell lines viz., human cancer cell lines A549 (lung carcinoma), MCF-7 (breast adenocarcinoma), DU145 (prostate carcinoma), DLD1 (colorectal adenocarcinoma) and FaDu (squamous cell carcinoma of pharynx) using SRB assay. Leaf and stem of CO are used for treating bone- and joint-related diseases by Indian traditional healers. The skeletal effects of CO and its phytochemicals at the cellular, molecular and organismal levels and the potential of CO extract in the treatment of skeletal diseases that lack effective therapy in mainstream medical practice are also discussed.

Introduction 1

The use of medicinal plants for healing purposes is as old as mankind. Several of the life-saving drugs in modern medicine originated from plants. Since very ancient times, humans have used plants as medicine for their wellness. The plant kingdom is a storehouse of a variety of organic components; many of these molecules are having excellent medicinal properties. Moreover, herbal medicines play an important role in the primary healthcare of the world's population. The main traditional systems of medicine practised in India are Ayurveda, Homeopathy, Siddha and Unani using plant-based medicines. Herbal formulations play an important role in the successful management of various diseases without any side effects. Indian medicinal plants are commonly used in home remedies against multiple ailments due to their pharmacologically active principles and compounds (Gupta, 2010).

Cassia species (*Caesalpinaceae*) have been well known for their laxative and purgative purposes. *Cassia* invites attention of researches worldwide for its phytochemistry and pharmacological activities. *Cassia* species are already reported in the ancient ayurvedic literatures for their use against various skin diseases such as ringworm, eczema and scabies. According to Ayurveda the leaves and seeds are acrid, laxative, antiperiodic, anthelmintic, ophthalmic, liver tonic, cardiotonic and expectorant. The leaves and seeds are useful in leprosy, ringworm, flatulence, colic, dyspepsia, constipation, cough, bronchitis, cardiac disorders. *Cassia* is an indigenous plant in Africa, Latin America, Northern Australia and Southeast Asia. Several *Cassia* species are of high commercial and medicinal significance due to their use in traditional medicines (Danish et al., 2011). The most widely recognized species of this genus are *Cassia auriculata*, *Cassia fistula,* and *Cassia occidentalis, Cassia siamea* and *Cassia uniflora* shown in Figure 1.1. *Cassia* species are reported to have various pharmacological activities such as antibacterial (Wadekar et al., 2011), analgesic, anti-inflammatory and antiarthritic (Manogaran and Sulochana, 2004; Nsonde et al., 2010; Chaudhari et al., 2012), hepatoprotective (Chauhan et al., 2009), antitumor (Gupta et al., 2000), antifertility (Yadav and Jain, 2009), antifungal (Singh and Karnwal, 2006), antioxidant (Siddhuraju et al., 2002; Danish et al., 2011), antileishmaniatic (Sartorelli

DOI: 10.1201/9781003186281-1

FIGURE 1.1 Images of selected plants of *Cassia*.

et al., 2007), antimicrobial (Ali et al., 1999), CNS (Mazumder et al., 1998) and hypoglycaemic activity (Bhakta et al., 2001; Pari and Latha 2002; Nirmala et al., 2008; Chauhan et al., 2009).

The taxonomy and nomenclature of *Cassia* species are quite complex. It is very difficult to differentiate them due to their overlapping morphological characters and close similarities (Kumar et al., 2007). This usually leads to misidentification and misinterpretation of the components. Different class of compounds reported from *Cassia* species are anthraquinones, phenolics, flavonoids, chromenes, terpenes, proanthocyanidins, coumarins, chromones, aromatic compounds and lignans (Lee et al., 2001; Danish et al., 2011; Zhao et al., 2013). Genus *Cassia* is in great demand due to its medicinal properties and newer formulations are continually appearing in the market (Chandra et al., 2015). The quality and safety of the products have to be checked regarding adulteration with misidentified species, which results in variations in phytoconstituents. It is therefore important to develop an efficient method, which will allow the discrimination in terms of distribution of bioactive compounds in different plants parts of *Cassia* species. Till now, a series of methods, such as high performance thin layer chromatography (HPTLC) (Bhope et al., 2010; Kadarkarai, 2011; Shailajan et al., 2013), multiwavelength high-performance liquid chromatography (Ni et al., 2009), high-performance liquid chromatography (HPLC) (Prakash et al., 2007; Ni et al., 2009; Lai et al., 2010; Chewchinda et al., 2012; Mehta, 2012; Sakulpanich et al., 2012; Chewchinda et al., 2013; Chewchinda et al., 2014), gas chromatography/mass spectrometry (GC-MS) (Tzakou et al., 2007), have been reported for the determination of anthraquinones, proanthocyanidins and/or phenolics. However, these methods suffer from low sensitivity, low resolution (Prakash et al., 2007; Ni et al., 2009; Chewchinda et al., 2012; Chewchinda et al., 2013; Chewchinda et al., 2014). There is a need to develop fast, selective and sensitive method for identification and determination of phytochemicals. Nevertheless, to the best of our knowledge, there is no report on the use of UPLC-QqQ_{LIT}-MS/MS technique for analyzing targeted compounds in different plant parts of *Cassia* species.

1.1 GEOGRAPHICAL DISTRIBUTION OF GENUS *CASSIA*

Cassia is a large genus of around 5000 species of flowering plants in the family *leguminaceae/fabaceae*. Description of selected *Cassia* species is reported in Table 1.1 (Hooker, 1975; Maitya et al., 1997; Harshal et al., 2011; Aditi et al., 2014; Awomukwu et al., 2015; Lavanya et al., 2018). Mostly *Cassia* species are annual shrub and grows all over the tropical countries (throughout India, Pakistan, Bangladesh and West-China) and in wasteland as a rainy season weed. It also grows in low lying coastal area, riverbanks, abundant in waste places and other moist places like uncultivated fields, up to 1000–1400 meters (Jain, 1968; Sanjivani, 2013; Lavanya et al., 2018). The selected species of *Cassia* such as *C. auriculata, C. occidentalis, C. siamea* and *C. fistula* are found throughout India are shown in Figure 1.2.

1.2 PHYTOGRAPHY OF GENUS *CASSIA*

Cassia species are wild crop and grown in most parts of India as a weed. It is an annual herb, 30–90 cm high. Leaves are green in colour, pinnate, up to 6–8 cm long, leaflets are in 3 pairs, distinctly petiole, opposite, conical at one end, ovate, oblong and base oblique. Flowers are pale yellow in colour usually in nearly sessile pairs in the axils of the leaves with five petals, upper one are very crowded. Pods are subteret or 4 angled, very slender, 6–12 inch long, incompletely septate, membranous with numerous brown oblong rhombohedral seeds (Maitya et al., 1997; Shivjeet et al., 2013).

TABLE 1.1 Selected species of *Cassia* and their medicinal properties. (Hooker, 1975; Maitya et al., 1997; Harshal et al., 2011; Aditi et al., 2014; Awomukwu et al., 2015; Lavanya et al., 2018)

S. No	*Name*	*Common Names*	*Phytochemical Constituents*	*Medicinal Uses*
1.	*C. fistula*	Hindi: Sonali, Amultus	Anthraquinones,	Anti diabetic activity
		English: Golden	Flavonoids,	Hypolipidemic activity
		Shower	Terpenoids,	Hepatoprotective activity
		Tamil: Shrakonnai	Reducing sugars,	Antioxidant activity
			Saponins,	Antipyretic activity
			Tannins,	Anti-inflammatory activity
			Carbonyl phlobatanin,	Antitussive activity
			Steroids,	Anti-laishmanial activity
			Glucoside,	CNS activity
			Rheinglucosides.	Antimicrobial activity
				Antimicrobial activity
				Antitumor activity
				Anti-ulcer activity
2.	*Cassia javanica*	Hindi: Javaniki-Rani	Anthraquinones,	Hypoglycemic activity
		English: Java *Cassia*	Reducing sugars,	Anticancer and antimycotic activity
		Tamil: Kondrai	Proteins,	Antioxidant activity
			Alkaloids,	Antiviral activity
			Tannins,	Antimicrobial activity
			Glycosides,	Haemolytic activity
			Flavonoids,	
			Sterols,	
			Quercetin,	
			Emodin	
			Chrysophanol,	
			Physcion.	

(*Continued*)

TABLE 1.1 (Continued)

S. No	Name	Common Names	Phytochemical Constituents	Medicinal Uses
3.	*Cassia grandis*	Pink Shower	Anthraquinones,	Anti-inflammatory activity
		Stinking Toe	Sterols,	Medicinal
		Coral Shower	Flavonoids,	Source of medicine
		Carao	Naphthalene derivatives,	Other Uses
			Protein,	Animal feed
			Tannins,	Ornamental purpose
			Alkaloids.	Revegetation
				Materials
				Gum, wood, timber
4.	*Cassia abbreviata*	Long pod *Cassia*	Anthraquinone derivatives,	Anti-plasmodic activity
			Guibourtinidiol,	Treatment for Malaria
			Alkaloids,	Treatment for Pneumonia
			Tannins,	
			Crude proteins,	
			Flavonoids,	
			Sterols.	
5.	*Cassia occidentalis*	Kasondi	Anthraquinone,	Treatment
			Anthrone,	Stomachic
			Cassiolein,	Flatulence
			Quercertin,	Constipation
			Aloe-emodin,	Cough
			Rhein,	Fever
			Tannins.	Asthma
6.	*Cassia obovata*	Neutral Henna	Anthraquinones,	Inhibitors of skin fungus
			Chrysophanic acid,	Mice infestations
			Tannins,	
			Sterols,	
			Flavonoids.	
7.	*Cassia spectablis*	Spectacular *Cassia*	Flavenol,	Antifungal activity

(*Continued*)

TABLE 1.1 (Continued)

S. No	*Name*	*Common Names*	*Phytochemical Constituents*	*Medicinal Uses*
			Anthraquinone, Tannins, Alkaloids, Emodin.	Antibacterial activity Antioxidant activity Antidiarrhoeal activity
8.	*Cassia tora*	Sickle Pod Thakara Coffee Pod Tovara	Cinnamaldehyde, Gum, Tannins, Mannitol,	Laxative Anthelminitic activity Ophthalmic use Antiperiodic

FIGURE 1.2 Distribution of selected *Cassia* species in India.

1.3 PHYTOCHEMISTRY OF GENUS *CASSIA*

Phytochemical screening of the plants and callus extracts employing TLC indicated glycoside, flavonoids, anthrone and anthracene derivatives (Table 1.2). It contains 1–2% volatile *Cassia* oil, which is mainly responsible for the spicy aroma taste. The primary chemical constituents of *Cassia* include cinnamaldehyde, gum, tannins, mannitol, coumarins and essential oils (aldehydes, eugenol and pinene); it also contains sugars, resins and mucilage, among other constituents (Singh et al., 1990; Shivjeet et al., 2013; Sanjivani, 2013; Lavanya et al., 2018).

1.3.1 Leaves

The leaves showed mainly the presence of Anthraquinone glycosides and Flavonoids. The Anthraquinone glycoside includes rhein, emodin, physion, chrysophanol (marker), obtusin, chryso-obtusin, chryso-obtusin-2-O-β-D-glucoside, obtusifolin and chryso-obtusifolin-2-O-β-D-glucoside (Singh et al., 1990; Shivjeet et al., 2013).

TABLE 1.2 Phytochemical constituents of *Cassia* species (Sanjivani, 2013; Lavanya et al., 2018)

Parts	*Chemical Constituents*
Leaves	Anthraquinone glycosides, rhein, emodine, physion, chrysophanol), Obtusin, chrysoobtusin, chryso-obtusin-2-O-β-D-glucoside,obtusifolin Flavnoids
Root	Betulinic acid, chrysophanol, Physcion, Stigmasterol, 1hydroxy-7-methoxy-3-methylanthraquinone, 8-O-methylchrysophanol, 1- Omethylchrysophanol, Aloe-emodin
Seed	Anthraquinones, Aurantio-obtusin, Chryso-obtusin, obtusin, Chryso-obtusin-2-O-β-D-glucoside, Physcion, Emodin, glycosides rubrofusarintri glucoside, nor-rubrofusarin gentiobioside, demethyl flavasperone gentiobioside, torachrysone gentiobioside, torachrysone tetraglucoside, torachrysone apioglucoside. Gums (7.65%)
Stem bark	Anthraquinones1hydroxy-5-methoxy-2-methyl anthraquinone, D-mannitol, myricyl alcohol, β-sitosterol, glucose, tigonelline, 1-stachydnine and choline.

1.3.2 Root

Eight compounds were isolated from the ethyl acetate fraction of *Cassia obtusifolia*, which are betulinic acid, chrysophanol, physcion, stigmasterol, 1-hydroxy-7-methoxy-3-methyl-anthraquinone, 8-O-methylchrysophanol, 1-O-methylchrysophanol and aloe-emodin (Singh et al., 1990; Shivjeet et al., 2013).

1.3.3 Stem Bark

The isolation of anthraquinone, 1-hydroxy-5-methoxy-2-methyl anthraquinone and its glycoside, 5-methoxy-2-methyl anthraquinone-1-O-α-L-rhamnoside along with chrysophanol, emodin and β-sitosterol from the stem of *Cassia* species is reported. The stem also contains D-mannitol, myricyl alcohol, β-sitosterol, glucose, tigonelline, 1-stachydnine and choline. The stem-bark yields ethyl arachidate and behenic acids, marginic and palmitic acids, euphol, aurapterol, basseol, rhein, 3, 5, 8, 3'4'5'-hexahydroxy flavones (Kapoor et al., 1980; Shivjeet et al., 2013).

1.3.4 Seed

Seed contains anthraquinones, namely; aurantio-obtusin, chryso-obtusin, obtusin, chrysoobtusin-2-O-β-D-glucoside, physcion, emodin, chrysophanol, obtusifolin, obtusifolin-2-O-β-D-glucoside, alaternin 2-O-β-D-glucopyranoside (Lee et al., 1998), brassinosteroids (brassinolide, castasterone, typhasterol, teasterone and 28-norcastasterone) and monoglycerides (monopalmitin and monoolein) (Park et al., 1994). Phenolic glycosides such as rubrofusarin triglucoside, nor-rubrofusarin gentiobioside, demethylflavasperone gentiobioside, torachrysone gentiobioside, torachrysone tetraglucoside and torachrysone apioglucoside were also isolated from the seed (Hatano et al., 1999). The seeds yield a gum (7.65%) which is the most efficient suspending agent for calomel, kaolin and talc. Extraction of the dried and crushed seeds with petroleum ether gave brownish-yellow oil and subsequently Chrysophanic acid was also isolated from this oil (Farooq et al., 1956). Thirteen phenolic glycoside including six new compounds were isolated from seed of *Cassia* species. These are rubrofusarin triglucoside, nor-rubrofusarin, gentiobioside, demethyflavasperone gentiobioside, torachrysone gentiobioside, torachrysone tetraglucoside and torachrysone apioglucoside. Two new naphtha-pyrone glycosides, 9 (β-D-glucopyranosynl-(1—6)-O-β-D-glucopyranosyl)oxy]-10-hydroxy-7-methoxy-3-mehtyl-1H-naptho[2,3-c]pyran-1-one and 6-O-β-D-glucopyranosyl)oxy]-rubrofusarin, together with *Cassia* side and rubrofusarin-6-β-gentiobioside were isolated from the seeds of *Cassia* species (Shivjeet et al., 2013).

1.4 PHARMACOLOGICAL ACTIVITIES OF GENUS *CASSIA*

The pharmacological activity profile of many *Cassia* species is of great importance. Some of them like *C. glauca C. angustifolia* and *C. acutifolia* are included in the pharmacopeia. Some of the important activities such as antioxidant, antidiabetic, hepatoprotective activity and cytotoxic activity along with many other activities of *Cassia* species are shown in Table 1.3 (Safaa et al., 2019).

1.5 TRADITIONAL USES OF GENUS *CASSIA*

Traditionally, the leaves of *Cassia* species are popular as *potherb*. It is used as a natural pesticide in the organic farms of India. It has been reported that *Cassia* species contain chrysophanic acid-9-anthrone which is an important fungicide. The intake of these seeds can cure skin diseases like ringworm, itch and psoriasis. These herbal seeds can also remove intense heat from the liver and improve the acuity of sight and loosen the bowels to relieve constipation. The leaves contain anthraquinones and are employed in weak decoction for treating childhood teething, fever and constipation. The paste of the ground, dried root is used in Ayurveda to treat ringworm and snakebite (Shivjeet et al., 2013).

1.6 INTRODUCTION OF SELECTED *CASSIA* SPECIES

1.6.1 *Cassia* Auriculata

1.6.1.1 Botany

C. auriculata Linn, family Leguminosae, is also known as *Avaram* tree, the leaves are alternate, stipulate, paripinnate compound, very numerous, closely

TABLE 1.3 Biological activities of *Cassia* species (Safaa et al., 2019)

Biological activities	*Species*	*Part or Compound Responsible for the Activity*
Antidiabetic	***C. italica***	Methanol leaves extract the ethanolic extracts of aerial parts
	C. javanica	Ethanol leaves extract
	C. nodosa	Stem & bark methanolic extracts - Emodin - Emodin 1-*O*-β-D-glucopyranosyl-(1→6)-*O*-β-D-glucopyranoside.
	C. roxburghii	- Aloe-emodin 8-*O*-β-D-glucopyranosyl-(1→6)-*O*-β-D-glucopyranoside. (leaves)
	C. siamea	Methanol leaves & flowers extract Flower extract
	C. semen	Leaves aqueous extract
	C. tora	Leaves extract Methanol extract
	C. alata	Leaves extract
	C. auriculata	Leaves aqueous extract Polyphenolic extract Leaves acetone extract Leaves aqueous extract Bark aqueous extract Flower aqueous and methanol
	C. glauca	extracts
Anti-ulcer activity		Ethanol leaves extract Polyphenolic extract
	C. semen	Leaves aqueous extract
	C. siamea	Leaves extract
	C. auriculata	Leaf extract
	C. alata	Kaempferol 3-*O*-sophoroside Leaves extract
	C. fistula	The ethanolic extract of the aerial
Analgesic and anti-inflammatory activity		part
	C. siamea	Stem bark extract
	C. uniflora	Leaves extract

(*Continued*)

TABLE 1.3 (Continued)

Biological activities	*Species*	*Part or Compound Responsible for the Activity*
	C. bakeriana	Leaves and bark ethanol extracts
		Leaves and flower extracts
		Seed methanol and acetone extracts
	C. glauca	Leaves methanol extract
Antibacterial		Flower methanol extract
		Seed extract
	C. nigricans	Ethyl acetate leaves extract
	C. occidentalis	Emodin (Root extract)
	C. reningera	Methanol flower extract
	C. sophera	Ethyl acetate seed coat extract
		Ethanolic leaves extract
	C. alata	Leaves extract
Antifungal		Leaves acetone extract
	C. glauca	Flower methanol extract
		Seed methanol and acetone extracts
	C. spectabilis	Flower methanol extract
	C. tora	Ethanolic leaves extract

placed, rachis 8.8–12.5 cm long, narrowly furrowed, slender, pubescent, with an erect linear gland between the leaflets of each pair, leaflets 16–24, very shortly stalked 2–2.5 cm long 1–1.3 cm broad, slightly overlapping, oval oblong, obtuse, at both ends, mucronate, glabrous or minutely downy, dull green, paler beneath, stipules very large, reniform-rotund, produced at base on side of next petiole into a filliform point and persistent (Prakash et al., 2014). Its flowers are irregular, bisexual, bright yellow and large (nearly 5 cm across), the pedicels glabrous and 2.5 cm long. The racemes are few-flowered, short, erect, and crowded in axils of upper leaves so as to form a large terminal inflorescence (leaves except stipules are suppressed at the upper nodes). The 5 sepals are distinct, imbricate, glabrous, concave, membranous and unequal, with the two outer ones much larger than the inner ones. The petals are free, imbricate and crisped along the margin, bright yellow veined with orange. The anthers are separate, with the three upper stamens barren; the ovary is superior, unilocular, with marginal ovules. The fruit is a short legume, 7.5–11 cm long, 1.5 cm broad, oblong, obtuse, tipped with long style base, flat, thin, papery, undulately crimpled, pilose, pale brown. 12–20 seeds per fruit are carried each in its separate cavity.

1.6.1.2 Medicinal Property

In Ayurvedic system of medicine, *C. auriculata* Linn. (Caesalpinaceae), commonly known as *Tanners Senna*, (Tarwar in Hindi) is a common plant used as antidiabetic, antidysentric, antimicrobial and for various skin diseases from ancient times. In Ayurveda, its seeds are used to treat various gastro-intestinal disorders. The flowers of the plant used as folk remedy for the treatment of Diabetes mellitus in southern parts of India. However, no scientific study on the antidiabetic property of *C. auriculata* plant has been reported. Its bark is used as an astringent, leaves and fruits as anthelminthic. The root cures tumors, skin diseases and asthma; leaves are anthelmintic, good for ulcers, diarrhea, and leprosy; and the flowers are used in the treatment of urinary discharge, diabetes, and dysentery. *C. auriculata* is also one of the major components of a beverage called "kalpa herbal tea" which has been widely consumed by people suffering from diabetes mellitus, constipation, and urinary tract diseases. An antidiabetic formulation called "avarai panchaga choornam" is prepared from dried and powdered plant parts (equal amount of leaves, roots, flowers, bark, and unripe fruits) are commonly used for opthalmia, con-junctivitis, and urinary infections also. The plant has been reported to possess antipyretic, hepatoprotective anti peroxidative and microbicidal activity. The flowers are used to treat urinary discharges, nocturnal emissions, diabetes and throat irritation. They are one of the constituent of polyherbal formulation. Decoction of the bark is effective in treating impotency and sexual related problems, helpful in avoiding leucorrhoea and per-vaginal discharge. The leaves are revitalizing as they alleviate feverishness and create a feeling of coolness and freshness. The hypolipidemic activities of the aqueous extract of *C. auriculata* flowers in streptozotocin induced diabetic condition suggest the beneficial effect against atherosclerosis. The *C. auriculata* flower extract has ability to regulate cholesterol, free fatty acids, triglycerides and phospholipids level and indicated antihyperlipidemic effect of the flower constituents. Fresh flowers of *C. auriculata* were extracted with ethanol and evaluated for antioxidant activities by 2, 2-Diphenyl 1-picryl hydrazyl solution (DPPH), 2,2'-azino-bis(3-ethylbenzthiazoline-6-sulphonic acid (ABTS) assay and anti-inflammatory activities by human blood cell (HRBC) membrane stabilization method and Inhibition of albumin denaturation method. It was observed that the ethanolic crude extract of *C. auriculata* flowers can be considered as good sources of antioxidants, anti-inflammatory and can be incorporated into the drug formulations.

1.6.1.3 Phytochemistry

The Bark of plant contains tannins, dimeric procyanidins, fisetinidol-(4, 8″)-catechin, fisetinidol-(4 8″)-epicatechin, fisetinidol-(4–8″)-gallocatechin, fisetinidol-(4- 8″)-epigallocatechin, rutin, 4′, 5, 7-trihydroxyflavan-3,4-diol, 5,2′,4′- trihydroxyflavan-3,4-diol. The heart wood depicts presence of anthocyanin glycoside, pelargonodin-5-O-D-galactoside 1, 3, 8-trihydroxy-2-methyl anthraquinone, 5-acetonyl-7-hydroxy-2-methylchromone. Leaves of plant contain 5-acetonyl-7-D-glucopyranosyl oxy-2-methylchromone, myristylalcohol, sitosterol-D-glucoside, quercetin3-O-glucoside, rutin and di-2-ethyl-hexylphthalate. Roots contains 1,3-dihydroxy-2-methylanthraquinone;1,3,8-trihydroxy-6-methoxy-2-methyl anthraquinone,1,8-dihydroxy-6-methoxy-2-methyl anthraquinone -3-Orutinoside, 1,8-dihydroxy -2-methyl anthraquinone-3-O-rutinoside, 4′-dihydroxyflavone-5-O-D-galactopyranoside. A chalcone 3′,6′- dihydroxy-4-methoxy chalcone, leucocyanidin-3-O-L-rhamnopyroside and leucopeonidin-3-O-L-rhamnopyroside were also present in root. Flowers contains auricassidin, kaempferol, α-sitosterol, octasonol, hentrioctanol, ceryl alcohol, peonidin-3-O-L-rhamnopyranoside, steroids, triterpenoids, lipids and flavonoids. Fruits contain cinnammic acid, nonacosane, emodin, nonacosane - 6-one, chrysophanol, and rubiadin, 1, 3-dihydroxy -6,8-dimethoxy-2-methylanthraquinone, Velutin-5- O-L-rhamnopyranoside glycosides, kaempferol-3-O-mannopyranosyl-(1,4)- O-D-glucopyranoside1, 5-dihydroxy-4,7-dimethoxy-2-methylanthraquinone-3-O-L-rham nopyranoside,1,3,6,7,8-pentahydroxy-4-methoxy-2-methyl anthraquinone, 3-hydroxy-6,8-dimethoxy2-methylanthraquinone-1-O-D-glucopyroside, 3-hydroxy-6,8-dimethoxy-2-methylanthraquinone-1-O-rhamnopyranoside, chrysophanol and physcion. The oil content in the seeds exhibits cholesterol, stigmasterol, sitosterol with palmitic, stearic, oleic & linoleic acid. Leaves and flowers have shown to contain anthraquinone glycosides, terpenoid glycosides, protoanthocyanidin, flavonol glucosides, hydroxyl anthraquinones and their glycosides, phenolic acids (Surveswaran et al., 2007). Seed has the presence of n-Hexadecanoic acid, 9-Octadecenoic acid, 1, 3, 12-Nonadecatriene and Stearic acid. The GC-MS analysis of ethanolic seed extract of *C. auriculata* revealed that the presence of benzoic acid, 2-hydroxy-, methyl ester (0.07%), glycine, N-(trifluoroacetyl)-, 1-methylbutyl ester (0.10%), 2,3-dihydro-3,5-dihydroxy-6-methyl-4Hpyran-4-one (0.12%), capric acid ethyl ester (0.16%), Resorcinol (0.21%) are showed as minimum percent. The fatty acid and fatty acid ester derivatives are recorded predominantly. Grape seed oil (Linoleic acid-21%, Oleic acid-7%, Palmitic acid-2.95%,) n-Hexadecanoic acid (21.31%), 9-Octadecenoic acid, (E)-(12.60%), Stearic acid (9.39%) and also the contribution of long-chained unsaturated hydrocarbon presents E,Z-1,3,12-Nonadecatriene (12.27%), dl-.α.-Tocopherol (1.22%), stigmasta-5, 23-dien-3-ol, (3.β.)-(1.21%).

1.6.2 *Cassia* Fistula

1.6.2.1 Botany

C. fistula L. (Leguminosae) a semi-wild Indian Laburnum (also known as the Golden shower) is distributed in various countries including Asia, South Africa, Mexico, China, West Indies, East Africa and Brazil (Rajan et al., 2001). The ornamental value of *C. fistula* is due to flowering crown from May to August. *C. fistula* is known as king of ornamental plants. It is a medium-sized tree growing to 10–20 m tall with fast growth. The leaves are deciduous or semi-evergreen, 15–60 cm long, pinnate with 3–8 pairs of leaflets, each leaflet 7–21 cm long and 4–9 cm broad. The flowers are produced in pendulous racemes 20–40 cm long, each flower 4–7 cm diameter with five yellow petals of equal size and shape. The fruit is a legume that is 30–60 cm long and 1.5–2.5 cm broad, with a pungent odour and containing several seeds. The striking long cylindrical pods produced from pretty yellow flowers. The fruits ripen (linear-cylindrical, 30 to 50 cm long, 1.5 to 1.7 cm in diameter, transversally septate, dark brown to black, and indehiscent when ripe) in the months of April and May. Outside its habitat, the season for flowering and fruit ripening varies. The seeds are obovate-ellipsoid, biconvex in cross-section, ventrally flattened, 7.5 to 10.0 mm long, 6.0 to 7.5 mm wide, and 2.5 to 3.0 mm thick. The seed coat is light brown, smooth, shiny and cartaceous with fracture lines. When stored at room temperature viability of seeds can be up to 1 year. To keep seeds live for longer period, they must be placed in sealed plastic, glass, or metal containers in cold chambers at a temperature of 5 to 6 °C. The seeds stored in a cold chamber after 1 year can still germinate. These stored seeds germinate because of the hard seed coat; seeds could be soaked in boiling water for 5 min before planting to stimulate germination.

1.6.2.2 Medicinal Property

According to the World Health Organization (WHO), traditional medical practices are an important part of the primary healthcare delivery system to a 3.5 billion people in the developing world. Most developing countries have adopted traditional medical practices as an integral part of their culture (World Health Organization 2007). *C. fistula* L. belongs to Fabaceae. The whole plant is used to treat diarrhea; seeds, flowers and fruits are used to treat skin diseases, fever, abdominal pain and leprosy by traditional people

(Perry, 1980). This plant is widely used by tribal people to treat various ailments including ringworm and other fungal skin infections (Rajan et al., 2001). It is used by Malaialis tribe in India to treat nasal infection (Perumal Samy et al., 1998). The pulp of the ripe fruit has a mild, pleasant purgative action and is also used as an anti-fungal drug (Kasuko and Nagayo, 1951). Indian people use the leaves to treat inflammation, the flowers as a purgative and the fruit as antiinflammatory, antipyretic, abortifacient, demulcent, purgative and refrigerant. The plant is good for chest complaints, eye ailments, flu, heart and liver ailments and rheumatism. In Ayurvedic system of medicine, this plant is used in haematemesis, pruritus, leucoderma, diabetes and many other ailments (Alam et al., 1990; Asolkar et al., 1992). *C. fistula* is a mild laxative that is suitable for children and pregnant women. It is also a purgative due to the wax aloin and has been used to treat many intestinal disorders like ulcers. The plant has a high therapeutic value and it exerts antipyretic and analgesic effects. *C. fistula* Linn is used as an anti-periodic agent in the treatment of rheumatism and the leaf extract is also used for its anti-tussive and wound healing properties. It has been seen that the fibre and mucilage content in different parts of *C. fistula* used in the treatment of hypercholesterolaemia. It has been reported to possess antitumor, hepatoprotective, antioxidant properties. In addition, its action on the central nervous and inhibitory effect on leukotriene biosynthesis has also been suggested. *C. fistula* plant organs are known to be an important source of secondary metabolites, notably phenolic compounds. There are many reports on the antioxidant properties shown by the extract of *C. fistula*. Luximon-Ramma et al. (2002) studied the antioxidant activities of phenolic, proanthocyanidin, and flavonoid component in extracts of *C. fistula*. The plant produces different phytochemicals are used as insecticides for killing larvae or adult mosquitoes or as repellents for protection against mosquito bites.

1.6.2.3 Phytochemistry

The cuticular wax of leaves contain hentriacontanoic, triacontanoic, nonacosanoic and heptacosanoic acids; anthraquinone, tannin, oxyanthraquinone, rhein, (-) epiafzelechin-3-Oglucoside, (-) epicatechin, procyanidin B2, biflavonoids, triflavonoids, rhein, rhein glucoside, sennoside A, sennoside B, chrysophanol, physcion (Wealth of India 1980). Sennoside A & B, chrysophanol, physcion, rhein and its glucoside, kaempferol, quercetin (-) epiafzelechin and its 3-0-(3-D glucopyranoside, (-) epicatechin, procyanidin B, isomers of epiafzelechin epicatechin, (2S) 7,4 dihydroxyflavan epiafzelechin and epiafzelechin, and gallic, protocatechuic acid, ellagic acid, citric, malic, succinic acids were obtained from the plant. Other content contains sugar, tannic matter, albuminous starch, oxalate of calcium, sugar, gum, astringent matter, gluten, colouring matter, water, proteins (19.94) and carbohydrates (26.30%); arginine, leucine, methionine, phenylalanine, tryptophan, aspartic and glutamic acids. A new dimeric proanthocyanidin C were isolated along with (-) epiafzelechin, (+) catechin, kaempferol, dihydrokaempferol and 1,8-dihydroxy-3- methylanthraquinone. Catechin, epicatechin, kaempferol, fistucacidin fistucacidin, leucocyanidin, leucopelargonidin trimer, rhein glycoside, hexacosanol lupeol, p-sitosterol. 5,7,3,4-tetrahydroxy-6,8 dimethoxyflavone-3-O-a-arbinopyranoside 5,7,4 trihydroxy-6,8, 3' trimethoxyflavone-3, rhamnosyl (l-»2)-0,p-D-glucopyranoside II and a xanthone glycoside vis 1,8 dihydroxy-3,7 dimethoxy xanthone-4-O-a-L rhamnosyl (1 -»2)-0-P-Dglucopy ranoside. Anthraquinone glycosides, sennosides A & B, rhein and its glucoside, barbaloin, aloin, formic acid, butyric acid and their ethyl esters, oxalic acid, pectin, tannin, sugar, gum, astringent matter, gluten, nonatetrac ontanone, 2-hentriacontanone,1,8- dihydroxy-3-anthraquinone carboxylic acid. Wood of the plant contains barbaloin, rhein, ieucoanthocyanidin, Fistacacidin, dihydrokaempferol, fistucacidin, Catechin, kacmpferol. Steam bark contain two flavonol glycosides, 5,7,3′,4′-tetrahydroxy 6, 8-dimethoxy flavone-3-O-α-arabinopyranoside ($C_{22}H_{22}O_{13}$), 5,7,4′-trihydroxy-6,8,3′-trimethoxyflavone-3-O-α-L-rhamnosyl (1→2)-O-β-D-glucopyranoside ($C_{30}H_{36}O_{18}$), α xanthone glycoside, 1,8-dihydroxy-3, 7 dimethoxyxanthone-4-O-α-L-rhamnosyl(1→2)-O-β-D glucopyranoside, lupeol, ß-sitosterol and hexacosanol. Flower contains rhein, kaempferol, leucopelar gonidin tetramer, ceryl alcohol, fistulin rhamnoside, aurantiamide acetaie, stigmosterol β sitosterol, β-D glucoside and ester methyl eugenol.

1.6.3 *Cassia* Occidentalis

C. occidentalis (Linn.) also known as *Senna occidentalis* is a weed of the leguminosae family. *C. occidentalis* (CO) is useed in Ayurvedic and ethnotraditional forms of medicine. CO has several local names such as "Kasaundi," "Kasondi," "Doddaagace," "Kaasaari," "Kasamarda," "Thagarai" (Khare 2007). CO is branched, rigid, and partially woody, and length varies between 2.5 and 5.0 feet with few lateral roots. Leaves are 15–30 cm long with variable shapes. This plant is distributed throughout the tropical and subtropical regions of Australia, Africa, Asia, and North America, South America and Oceania (Liogier 1994; Stevens 2001). CO is widely available in India and grows during rainy season, completes flowering cycle in autumn and usually dies during winter. Harvest time of CO is generally considered to be right after flowering. There are several excellent reviews on the pharmacology of the extracts of CO (Yadav et al., 2010; H. Singh et al., 2019).

Although the traditional use of CO is for the healing of fracture and bone injuries, it could also impact other major skeletal pathologies such as metabolic and autoimmune destruction of bones (Yadav et al., 2010). Metabolic causes of bone loss include low levels of gonadal hormones (age-related) and excess of corticosteroid, parathyroid hormone, and thyroid hormones (Cannarella et al., 2019). Besides, a large population suffers from iatrogenic (drug-induced) osteoporosis most frequently caused by the chronic uses of glucocorticoids, aromatase inhibitors, proton pump inhibitors, anti-seizure medications, and heparin (Mirza and Canalis, 2015). Autoimmune destruction of joints, the prevalent form is Rheumatoid arthritis (RA), has a global prevalence of ~20 million, with an age-standardized prevalence rate of 246.6 per 100000 (Cowen, 2019). Osteoarthritis (OA), an age-associated disease caused due to the failure of chondrocytes to maintain the balance between synthesis and degradation of extracellular matrix components of articular cartilage, subchondral bones, and ligaments of the joints (Man and Mologhianu, 2014). Approximately 10–15% of those over 60 years have some degree of OA, with prevalence higher among women than men. Although there are many other diseases, age-related osteoporosis, iatrogenic osteoporosis, RA and OA are the most prevalent forms of the diseases of bone and joint.

1.6.3.1 Botany

C. occidentalis Linn usually grows in the southern part of India which is known as Kasmard in Sanskrit, Kasondi in Hindi and Coffee Senna in English. The plant belongs to Caesalpiniaceae family (Yadav et al., 2010; H. Singh et al., 2019). The common name is Ponnavarai in Tamil. It is an erect herb, commonly found by road sides, ditches and waste dumping sites. CO has been widely used as traditional medicine. Entire parts of the plant have medicinal values (Yadav et al., 2010; H. Singh et al., 2019). CO is an erect, somewhat branched, smooth, semi-woody, fetid herb or shrub, 0.8–1.5 m tall, tap root, hard, stout, with a few lateral roots on mid section. This plant species varies from a semi-woody annual herb in warm temperate areas to a woody annual shrub or sometimes a short-lived perennial shrub in frost-free areas. The stem of the plant is reddish-purple. The young ones are 4-sided, becoming rounded with age. Leaves are alternate, even pinnately compound, each one with 4–6 pairs of nearly sessile, opposite leaflets, with a fetid smell when crushed, each leaflet 4–6 cm long, 1.5–2.5 cm wide, ovate or oblong, lanceolate with a pointed tip and fine white hairs on the margin. The rachis has a large, ovoid, shining, dark purple gland at the base. Stipules are 5–10 mm long, often leaving an oblique scar. Inflorescence is a compound of axillary and terminal racemes. Flowers are yellow in color in short racemes, blooming in the month of July to August. The flower is perfect, 2 cm long with 5 yellowish-green sepals with distinct red veins and 5 yellow petals. The fruit is a dry, dehiscent, transversely partitioned, faintly recurved, laterally compressed, sickle shaped legume (pod), 7–12 cm long, 8–10 mm wide, with rounded tip and containing 25–50 seeds. Pods are recurved, glabrous and compressed. These are 10–12.5 cm long and 5mm. thick. Seeds are 20 to 30 in number, 6 mm long and 4 mm broad, dark olive-green in colour, ovoid, hard, smooth and shiny in appearance. Seeds are compressed at one end and rounded at other end. Seeds are oval-shaped 3.5–4.5 mm wide, flattened; pale to dark brown, slightly shiny, smooth and with a round pointed tip. This plant is widely consumed by animals and humans. However, some toxicological effects of seeds and leaves of this plant have been observed. Still, this plant is widely consumed by the local people as a coffee substitute. It is a main ingredient of Liv-52, a hepatoprotective polyherbal formulation.

1.6.3.2 Medicinal Property

Based on the traditional knowledge, CO has antimicrobial property. A test was conducted with four different CO extracts in methanol, aqueous, benzene, petroleum ether and chloroform extract against *P.aeruginosa, K. pneumoniae, P. mirabilis, E. coli, S. aureus* and *S. epidermidis*. Among the four different extracts the methanolic extract showed a better response against all the bacterial strain (Kumar et al., 2013). Another study reported maximum activity against *Salmonella typhi* and minimum with *Shigella* spp. (Das et al., 2010). CO flower extract showed maximum inhibition against *Klebsiella pneumonia* (Arya et al., 2010). The metabolite rich MeOH fraction of CO (anthraquinones) leaves, pods, flowers and callus were effective against *E. coli* also. Anti-inflammatory efficiency of CO was evaluated by delayed type hyper sensitivity method using mice which confirmed the anti-inflammatory effect. The wound healing property of crude extract of CO leaves and a pure compound chrysophenol isolated from it was evaluated in excision and dead space wound models. Chrysophenol from CO was found to possess significant wound healing property (Sheeba et al., 2009). The aqueous–ethanolic extract of leaves of *C. occidentalis* was tested for hepatoprotective activity on liver damage in rat which was induced by paracetamol and ethyl alcohol by monitoring serum transaminase, alkaline phosphatase, serum cholesterol, serum total lipids and histopathological alterations. The CO leaf extract had shown significant hepatoprotective activity. *C. occidentalis* is an ingredient in Himoliv, a polyherbal ayurvedic formulation. It prevents the carbon tetra chloride induced hepatotoxicity in rats (Dhanasekaran et al., 2009). CO plant extract has effective antimalarial activity (Tona et al., 2001). A study with ethanolic, dichloromethane and lyophilized aqueous extracts of *C. occidentalis* root bark was tested for antimalarial activity and found active against

Plasmodium berghei (Tona et al., 1999). *In-vitro* cytotoxicity properties of *C. occidentalis* plant via alcoholic, hydro alcoholic and aqueous extract was evaluated against eight human cancer cell lines from six different tissues in which aqueous extract showed maximum activity against six cell lines in dose dependant manner (Bhagat et al., 2010).

1.6.3.3 Phytochemistry

Phytochemical analysis of *C. occidentalis* showed qualitative and quantitative phytochemical variation according to climate. For example stems, leaves and the root bark of the plant from Ivory Coast and Africa contain a small amount of saponins, no alkaloids, sterols, triterpenes, quinines, tannins and flavonoids. However, a large amount of alkaloids were found in the stem, leaves and fruits from Ethiopia. Phytochemical screening of the plant showed the presence of carbohydrates, saponins, sterols, flavonoids, resins, alkaloids, terpenes, anthraquinones, glycoside, balsam, eudesmane sesquiterpenes and anthraquinones, emodin and chrysophanone. *C. occidentalis* root samples have been reported to possess 4.5% total anthaquinones. Emodin, 1,8-dihydroxy anthraquinone and the flavonoid quercetin were also identified. Young root samples have been found to possess chrysophanol and emodin. Physcion (free, bonded, reduced and oxidized) together with chrysophanol have also been reported. Along with these known anthraquinones, two xanthone derivatives pinselin and pigment E have also been observed. Two new bis (tetrahydro) anthracene derivatives, occidentalol-I and occidentalol-II were isolated from the roots of *C. occidentalis* along with chrysophanol, emodin, pinselin, questin, germichrysone, methyl germitorosone and singueanol-I. Two sterols named β-sitosterol and campesterol which usually occur together in a plant were also found in this plant. From the callus of *C. occidentalis* six anthraquinones—islandicine, chrysophanol, physcion, emodin, questin and 7-methyl-physcion, the bianthraquinones, chrysophanol, bianthrone, three tetrahydro anthracenes—germichrysone, methylgermitorosone and 7-methyltorosachrysone plus the xanthone pinselin were isolated (Yadav et al., 2010).

In the seed extract of *C. occidentalis* chrysophanol, toxins, 1,4-oxazine derivative n-methyl morpholine has been isolated. Seeds also contain physcion, physciondianthron heterosides and physcion condensed as homodianthrone as well as a mixture of anthraquinones. 1-glucoside of physcion (0.018%) along with physcion (0.0068%) and two new anthraquinones like 1, 8-dihydroxy-2-methyl anthraquinone (0.0014%) and 1,4,5-trihydroxy-3-methyl-7-methoxy anthraquinone (0.0016%), chrysophanol free and as a glycoside were also found in seed samples. A new polysaccharide galactomannan consisting of D-galactose and D-mannose in the proportion of 1:3.1, as well as trace amount of D-xylose was also found in the *C. occidentalis* seeds. Maltose, lactose, sucrose and raffinose are also detected from seed. Some other compounds identified from the seeds of *C. occidentalis* are -1,8-dihydroxy-2-methyl anthraquinone, physcion, rhein, aloe-emodin, chrysophanol and steroidal glucosides. An analysis of flowers indicated the presence of anthraquinones, emodin, physcion and physcion-1-O-β-D-glucoside as well as the ubiquitous sterol β-sitosterol (Veerachari and Bopaiah, 2012).

Two flavonoid glycosides 3,5,3′,4′- tetrahydroxy-7-methoxy flavone-3-O-(2′-rhamnosyl glucoside) (rhamnetin-neohesperidoside) (I) and 5,7,4′-trihydroxy- 3,6,3′-trimethoxy flavone 7-O-(2-rhamnosylglucoside) (II) have been isolated from the flower of CO. 1,8-dihydroxy-2-methyl anthraquinone; 1,4,5-trihydroxy-7-methoxy-3-methyl anthraquinone, physcion, rhein, aloe-emodin, chrysophanol and steroidal glycosides were also reported from pods of *C. occidentalis* plant. Other compounds reported in literature include, 1,8-dihydroxyl-2-methyl anthraquinone, 1,4,5- trihydroxy-3-methyl-7-methoxy anthraquinone, cassiaoccidentalin A, B and C, which are C-glycosides, achrosine, anthrones, apigenin, aurantiobtusin, campesterol, cassiollin, chryso-obtusin, chrysophanic acid, chrysarobin, chrysoeriol, essential oils, funiculosin, galactopyranosyl, helminthosporin, islandicin, kaempferol, lignoceric acid, linoleic acid, linolenic acid, mannitol, mannopyranosyl, matteucinol, obtusifolin, obtusin, oleic acid, physcion, quercetin, rhamnosides, rhein, rubrofusarin, sitosterols, and xanthorin. Overall phytochemical investigation of CO showed the presence of different classes of compounds including flavonoids, saponins, anthraquinones, steroids, alkaloids, terpenes, glycosides, sterols, and resins from different parts of the plant (Hatano et al., 1999). Major skeletal effects of these compounds are given in Table 1.4 and individually discussed in detail in sections 3.4. Several of these compounds have been pharmacologically investigated and their pharmacological effects are also summarized in Table 1.5.

1.6.4 *Cassia* Siamea

1.6.4.1 Botany

C. siamea was initially classified in family Caesalpiniacae, is now classified in family Fabaceae *C. siamea* is customarily planted in lanes with corn and cotton because the foliage is rich in organic matter and serves as green manure. Because it grows fast, the species is planted in wet tropical regions to produce firewood (Kamagaté et al., 2014). The wood is used for poles, turned articles, furniture, and pulp for paper and in rural construction. The bark contains tannin and is used to tan hides. The flowers, rich in nectar, is honey

TABLE 1.4 Compounds isolated from *C. occidentalis* and their skeletal effects

S. No.	*Plant Part*	*Name of the Compound*	*Structure*	*Biological Activity*	*Reference (PMID)*
1.	Whole plant	Quercetin	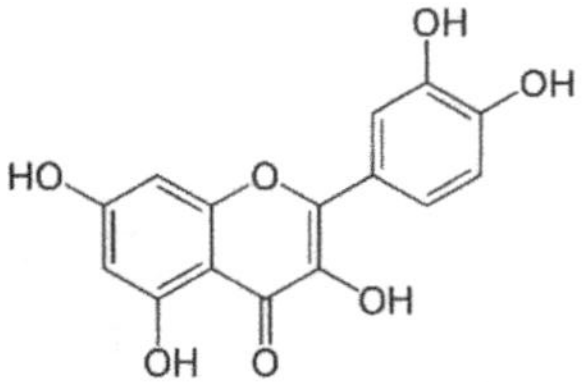	Anticancer	31680350
2.	Roots	Emodin	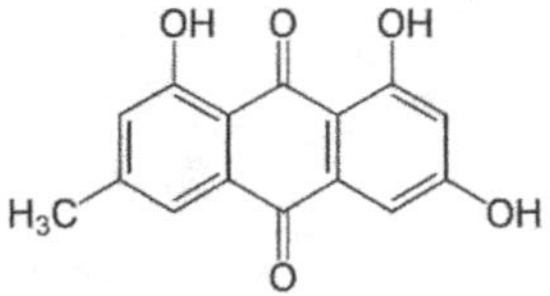	Osteogenic, Antibacterial, Anti-inflammatory, Laxative	30703497 , 31680350, 25026715, 24727085
		Cassiollin	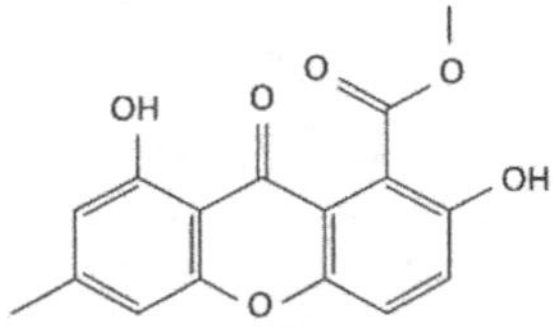	Antiplatelet, Anti-inflammatory	19796670

(*Continued*)

TABLE 1.4 (Continued)

S. No.	*Plant Part*	*Name of the Compound*	*Structure*	*Biological Activity*	*Reference (PMID)*
3.	Fruits	Rhamnetin	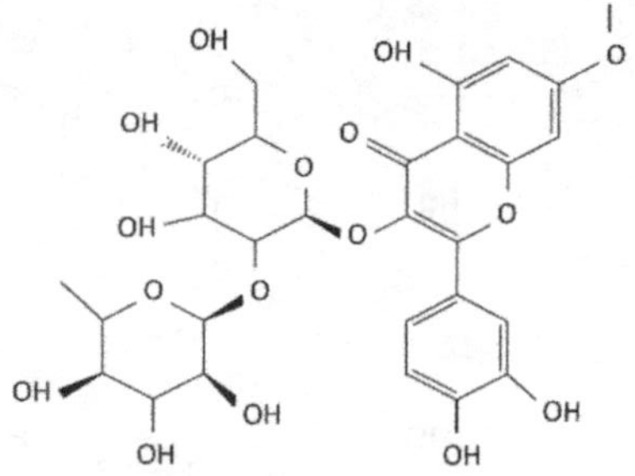	Antioxidant, Antidiabetic	17345279
4.	Leaves	2′,4′,5′,7′-tetrahydroxyflavanol	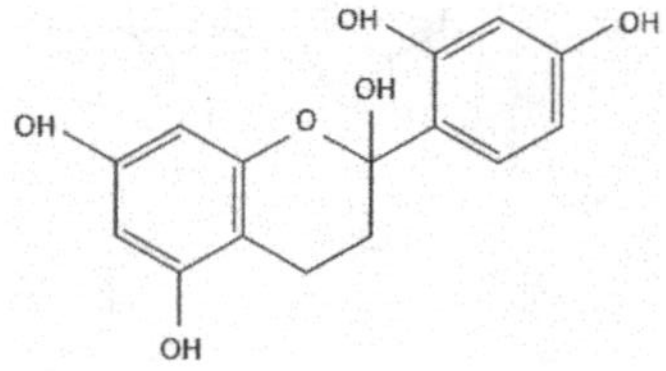	Anti-inflammatory, Anti-ulcer	29801717

Apigenin	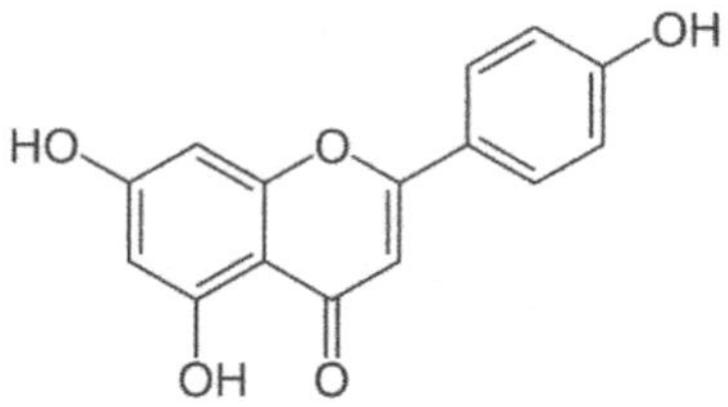	osteogenic	30703497, 31932603
Isovitexin	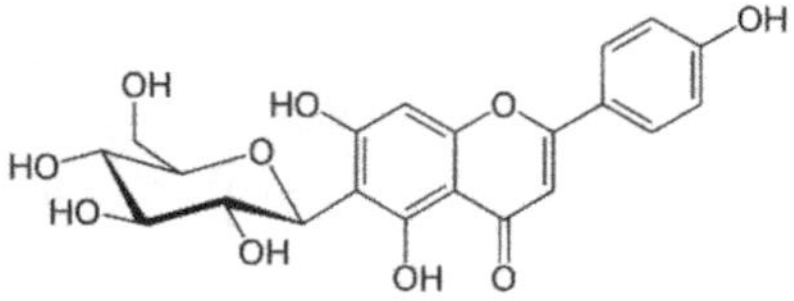	osteoanabolic	30703497, 31932603
Luteolin	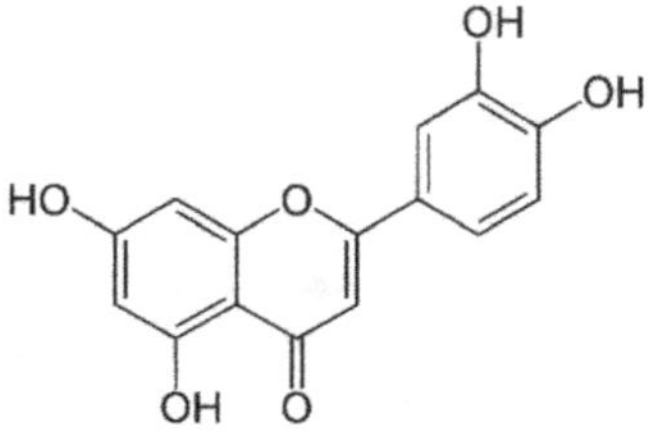	osteogenic	30703497
Chrysophanol	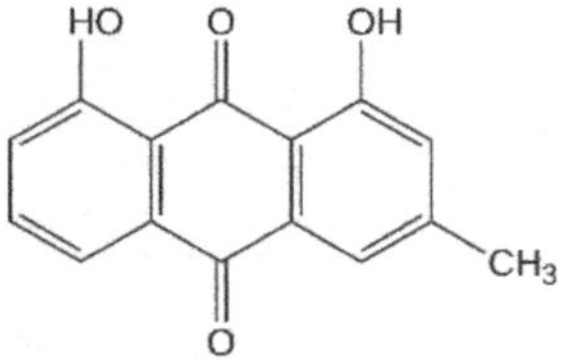	Anti-inflammatory,wound healing,	25026715, 19796670

(Continued)

TABLE 1.4 (Continued)

S. No.	*Plant Part*	*Name of the Compound*	*Structure*	*Biological Activity*	*Reference (PMID)*
		Kaempferol	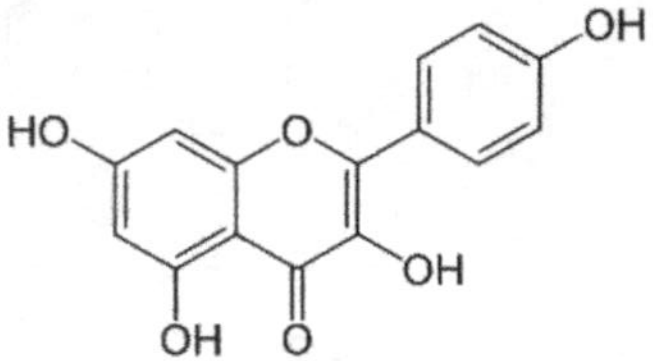	Osteogenic, anticancer	31678747, 29655780, 31638215 31248102
5.	Seeds	β-sitosterol	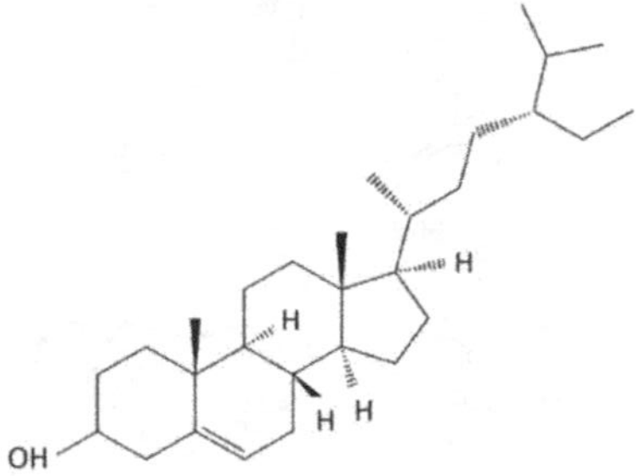	Hypolipidemic	20523870,

		Chrysophanol	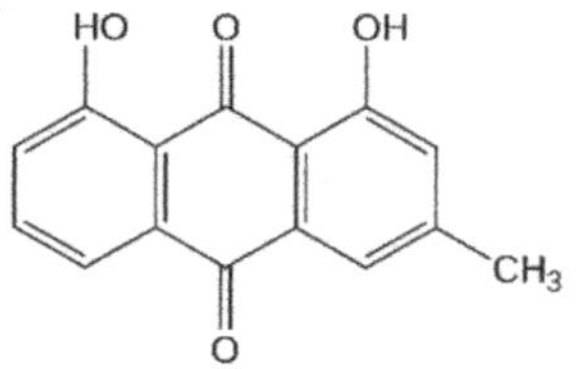 HO O OH CH$_3$ O	Anti-inflammatory, wound healing,	25026715, 19796670
6.	Bark	Chrysophanol	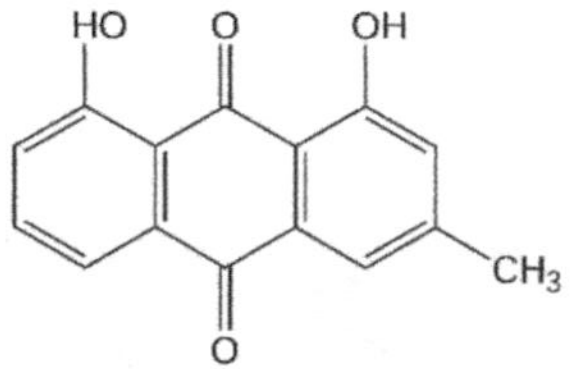 HO O OH CH$_3$ O	Anti-inflammatory, wound healing,	25026715, 19796670
		Physcion	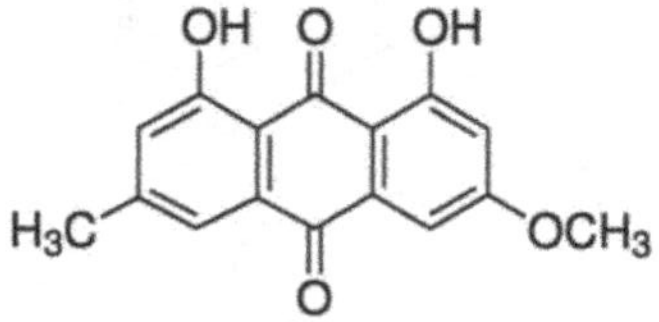 OH O OH H_3C OCH_3 O	Anti-cancer	25026715, 29710489
		Emodin	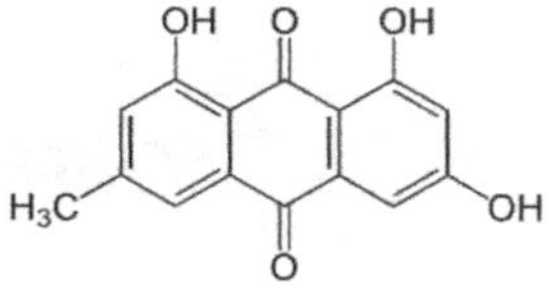 OH O OH H_3C OH O	Osteogenic, Anti-bacterial, Anti-inflammatory, Laxative	30703497, 31680350, 25026715, 24727085

(Continued)

TABLE 1.4 (Continued)

S. No.	*Plant Part*	*Name of the Compound*	*Structure*	*Biological Activity*	*Reference (PMID)*
7.	Stem	Apigenin	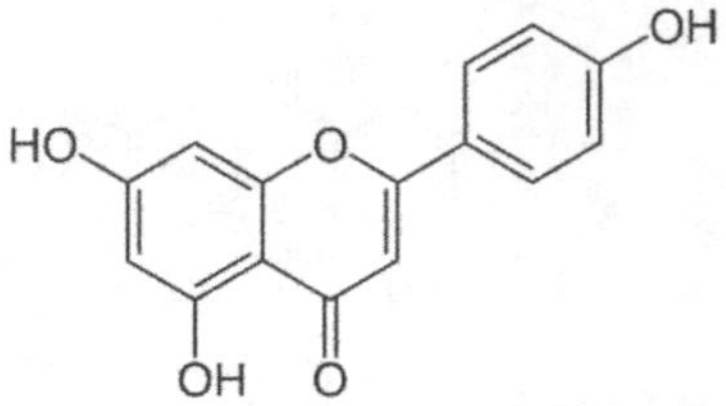 HO, OH, O, OH, O	osteogenic	30703497, 31932603
		4′,7′-dihydroxyflavone	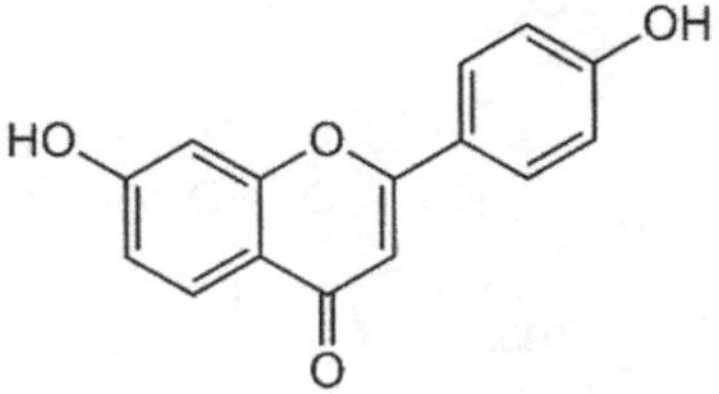 HO, O, OH, O	osteogenic	30703497
		Luteolin	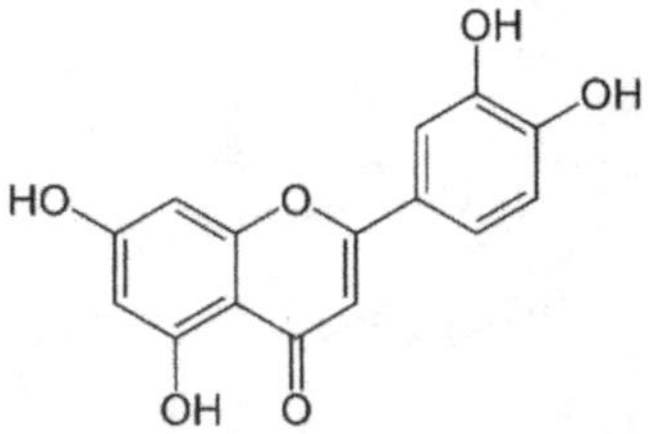 HO, O, OH, OH, OH, O	osteogenic	30703497

3′,4′,7′-trihydroxyflavone	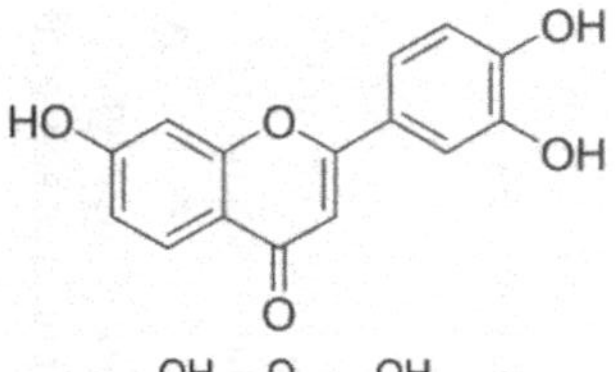	osteogenic	30703497
Emodin	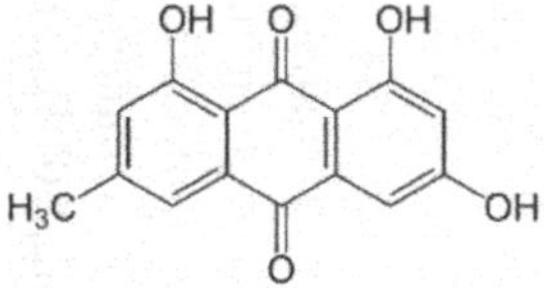	Osteogenic, Antibacterial, Anti-inflammatory, Laxative	30703497, 31680350, 25026715, 24727085
Isovitexin	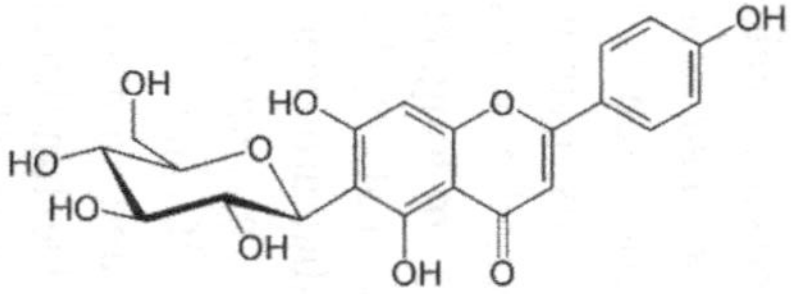	osteoanabolic	30703497, 31932603

TABLE 1.5 Pharmacologically investigated compounds of *C. occidentalis*

Part	*Compounds*
Seed	Aurantio-obtusin, 1,4,11,12-tetrahydro-9,10-anthraquinone, 1, 8-dihydroxy-2-methylanthraquinone, isochrysophanol, 5, 7'-biphyscion
Callus	Germichrysone, cassiolin, 7-methylphyscion, 7-methyltorosachrysone
Root	Xanthorin, helminthosporin, chrysophanol, 1,7-dihydroxy-3-methylxanthone, islandicin, 8-*O*-methylchrysophanol, 1-hydroxy-9,10-anthraquinone, diacerein,1,8-dihydroxyanthraquinone, methylgermitorosone
Leaf	Singueanol-I, occidentalol-1, occidentalol-II, chrysophanol-10, 10'-bianthrone,

bearing. The foliage, fruits, and seeds are fatal to pigs, but cattle and sheep are not affected by their toxicity. *C. siamea* is a fast-growing, short-lived and evergreen tree. Under optimal conditions, it can reach 30 m in height (Kamagaté et al., 2014). The tree has a straight trunk and a rounded or irregular and spreading, multi-branched crown with dense foliage. The leaves of *C. siamea* are pinnate, 23 to 33 cm long, and made up of 5 to 14 pairs of lanceolate, oblong or ovate elliptic leaflets, 3 to 7 cm long and 12 to 20 mm wide. The species requires soils that are deep, well-drained, and rich in organic matter for good development, its requires average annual temperature is 24.2 °C, with a minimum of 19.9 °C and a maximum of 27.7 °C and a dry season that lasts 4 to 6 months and rain in the summer. The tree grows naturally from sea level to 600 m (Deshmukh et al., 2014). It endures seasonal flooding, salinity, and continuous exposure to wind and shade. However, it is not very resistant to cold and drought. In its native habitat, *C. siamea* blooms precociously and abundantly from June to January. Outside its area of natural distribution, the tree blooms and fruits at different times of the year, depending upon the environment. The flowers have yellow petals and are arranged in racemes or panicles. *C. siamea* begins to fruit at 5 years. The fruits are hanging, linear, plano-compressed legumes, 5 to 30 cm long, 12 to 20 mm wide, and bicarinate, coriaceous or subwoody, and dark brown and dehiscent when ripe. Each fruit contains approximately 25 seeds (Deshmukh et al., 2014). The seeds ranges in shape from circular to obovate and in some cases are vaguely elliptic and laterally flattened. Seed size ranges from 6.5 to 8 or 9 mm long, 5.5 to 6.0 mm wide, and 0.8 to 1.0 mm thick. The seedcoat is dark brown, smooth, shiny, and cartaceous, and 3.3 to 4.5 mm long by 0.9 to 1.2 mm wide, with a closed, oblong-elliptic pleurogram on each of its lateral surfaces. Leves are evergreen tree, 10–12 m tall. Leaves pinnately compound

10–25 cm long, rachises pubescent. Leaflets 6–14 pair's ovate oblong, in *C. siamea* (lam.). Very few stomata are present on abaxial surface and large number of stomata is present on adaxial surface. Trichomes are found along the margin of abaxial surface and on adaxial surface trichomes are present on midrib and along the leaflets margin. Peripinnate about 12 inches long, leaflets 12 to 20, elliptical oblong, mucronate, glabrous. Leaves alternate, pinnately compound, 23–33 cm long, with slender, green reddish, tinged axis; leaflets 6–12 pairs on short stalks of 3 mm, oblong, 3–7 cm long, 12–20 mm wide, rounded at both ends, with tiny bristle tip. Bark is medium size, quick-growing tree to 12 m tall (40 ft) with relatively smooth bark, A straight trunk of up to 30 cm in diameter; the bole usually short, the crown usually dense and rounded at first, later becoming irregular and spreading with drooping branches. Bark grey or light brown, smooth but becoming slightly fissured with age (Singh et al., 2013). Flowers are yellow grow in large, open cluster at the ends of the branches about 11/4 inches, each of the flower having five almost equal petals and perfect seven stamens nearly unequal that produce pollen the remaining three stamens being waiting, or small and sterile. Flower clusters are upright at ends of twigs, large branched, 20–30 cm long, 13 cm broad, with many bright yellow flowers 3 cm across, pentamerous; sepals imbricate, obtuse at the apex; petals subequal to heteromorphic, yellow; stamens, accrescent toward the abaxial side of the flower; filaments straight and not more than twice as long as the anthers; ovary superior, linear and curved (Kamagaté et al., 2014).

1.6.4.2 Medicinal Property

Different parts of *C. siamea* can be used for various medical purposes. The leaves, stems, roots, flowers and seeds of *C. siamea* regardless of the sub-species have been used for the treatment of several illnesses including mostly malaria, a tropical endemic disease with high morbidity and mortality. The leaves are the most used parts of the plant mainly by African and Asian population in preparation of the herbal remedies. In Burkina Faso, fresh and dried leaves decoction (boiled for 20 min in 1L of water) is drunk with lemon juice or for body bath throughout the day to treat malaria and liver disorders. *C. siamea* has been reported to be used in the management of constipation, diabetes, insomnia, hypertension, asthma, typhoid fever, and dieresis. Leaves and bark of medicinal plants were reported to be used locally as antimalarial medications. Traditionally *C. siamea* is used for the treatment of typhoid fever, jaundice, abdominal pain, menstrual pain and is also used to reduce sugar level in the blood (Kamagaté et al., 2014). Ethno medicinally *C. siamea* is used as laxative, blood cleaning agent, cure for digestive system and genitourinary disorders, herpes and rhinitis. In traditional medicine, the fruit is used to charm away intestinal worms and to prevent convulsion in children. Among all the tested *C. siamea* leaves extracts, hexane extract have shown significant antibacterial activity. In Nigeria, the dried leaves are mixed with lemon's leaves (*Cymbopogon citratus*), pawpaw's leaves (*Carica papaya*), and the lime's leaves (*Citrus lemonum*) and are boiled within an hour. The "tea" of the mixture is drunk against malaria. In Uganda, the leaves are picked, cleaned and chewed, and liquid swallowed to treat abdominal pains. In India, the boiled leaves are used against anaemia and fever. The decoction of the flowers is used in body bath against malaria and liver disorders. This decoction is also effective against insomnia and asthma. In Sri Lanka and Thailand, the flowers and young fruits are regularly consumed as vegetable and for treating curries. It provides laxative and sleeping-aid effect. This dish is also anxiolitic and effective against dysuria. The dried stems of *C. siamea* mixed with the fruit of *Xylopia aethiopica* is pulverized and administered as laxative. The decoction of the stem bark is drunk against diabetes. This decoction is used as a mild, pleasant, safe, and purgative in Japan. The decoction is used against scabies, urogenital diseases, herpes, and rhinitis in Cambodia. The leaves and flowers of *C. siamea* are active on central nervous system. Anxiolitic effect of aqueous extracts of leaves and flowers were observed using an elevated plus-maze (EPM) test in rats. The *in-vitro* studies of antioxidant property showed that various extracts of *C. siamea* possessed high antioxidant potential measuring using β-carotene bleaching method, 2,2′-azinobis (3-ethylbenzothiazoline-6-sulfonic acid) (ABTS) radical cation and superoxide anion radical scavenging assay (Wagner et al., 1978; Singh et al., 2013; Kamagaté et al., 2014).

1.6.4.3 Phytochemistry

C. siamea contains different phytochemical compounds like lupeol, chrysophanol, cassiamin A, cassiamin, siameadin, lupeone, rhein, chrysophanolantrone, barakol, *cassia* chromone (5-acetonyl-7-hydroxy-2- methylchromone), p-coumaric acid, apigenin-7-O-galactoside, β-sitosterol, *cassia* chromonone and cassiadinine (Padumanonda et al., 2006). Leaves of plant contain anhydrobarakol; 5-acetonyl-7-hydroxy-2-methylchromone; 5-acetonyl-7-hydroxy-5-acetonyl-7-hydroxy-2- hydroxymethylchromone, barakol, cassiarin A, cassiarin B, chrysophanol, emodin, physion, rhein, sennosides, cassiamin A, cassiamin B, chrobisiamone A, resins, lupeol, D-pinitol, luteolin, 4-(trans)-acetyl-3,6,8-trihydroxy-3-methyldihydronaphthalenone; 4-(cis)-acetyl-3,6,8-trihydroxy-3-methyldihydro-naphthalenone, β- and γ-sitosterol, carotenes, xanthophylls, vitamin A, C, E, 2',4',5,7-tetrahydroxy-8-C-glucosylisoflavone. Stem bark contain 4-4'-bis(1,3-dihydroxy-2-methyl-6,8-dimethoxy-anthraquinone; 1,1'-bis(4,5-dihydsroxy-2-methyl-anthraquinone, cassiamin A, cassiamim B, cassiamin C; madagascarin, chrysophanol, emodin, physcion, chrysophanol-1-O-β-D-glucopyranoside; 1-[(β-D-glucopyranosyl-(1–6)-O-β-D-glucopyranosyl)-oxy]-8-hydroxy-3-methyl-9, 10-anthraquinone; cycloart-25-en-3β, 24-diol, piceatannol, 24-dihydroxyurs-12-ene-28-oicacid- 3-O-β-D-xylopyranoside, lupeol, friedelin, betulinic acid, coumarin, siamchromones A-G, 2-methyl-5-(2′(hydroxypropyl)-7-hydroxy-chromone-2′-O-β-D gluco-pyranoside; 2-methyl-5-propyl-7,12-dihydroxy-chromone12-O-β-D-glucopy-ranoside kaempferol. Root bark contain chrysophanol; emodin, 1,1′,3,8,8′-pentahydroxy-3′,6-dimethyl[2,2′-bianthracene] 9,9′,10,10′-tetrone; 7-chloro-1,1′,6,8,8′-pentahydroxy-3,3′-dimethyl[2,2′-bianthracene]-9,9′,10,10′-tetrone; cassiamin A, cassiamim B. Flowers contain barakol, 11-dihydroanhydro-barakol, cassiarin C, D, E, and F, cassiadinine, gallic acid; protocatechuic; p-hydroxy benzoic acid; chorogenic acid; vanilic acid; caffeic acid; syringic acid; p-coumaric acid; ferulic acid; sinapic acid, rutin; myricetin; quercetin; kaempferol. Seeds contain cholesterol, stigmasterol, β-sitosterol, palmitic, stearic, oleic and linoleic acids, aloe-emodin, sennosides A, vitamin B1, B2, B3, C, E, lysine, valine, leucine, isoleucine, threonine, methionine, cystine, tyrosine, histidine, arginine, aspartic acid, serine, glutamic acid, proline, glycine; alanine (Wagner et al., 1978; Parveen et al., 1995; Padumanonda et al., 2006; Odason et al., 2007).

2

Qualitative and Quantitative Determination of Bioactive Phytochemicals in Selected *Cassia* Species Using HPLC-ESI-QTOF-MS and UPLC-ESI-QqQ_{LIT}-MS/MS

DOI: 10.1201/9781003186281-2

2.1 PLANT MATERIAL AND CHEMICALS

Plant materials (leaf, stem, root, seed and flower of *Cassia auriculata, Cassia occidentalis, Cassia siamea, Cassia uniflora,* with fruit pulp instead of flower in case of *Cassia fistula*) were collected from the plants grown in western ghats of Maharashtra region between the months of May to September 2013. Voucher specimens of *C. auriculata* (ARC-5), *C. fistula* (ARC-2), *C. occidentalis* (ARC-3), *C. siamea* (ARC-1) and *C. uniflora* (ARC-4) were deposited in the Botanical Survey of India, Western Region Centre, Pune, India.

Acetonitrile, Methanol (LC-MS grade) and formic acid (analytical grade) were purchased from Fluka, Sigma-Aldrich (St. Louis, MO, USA) and ultra-high purity water was prepared using a Milli-Q water purification system (Millipore Corporation, Bedford, MA). Syringe filters (0.22 μm) were purchased from Millipore (Billerica, MA, USA). The reference standards (purity≥96%) of chrysophanic acid, physcion, rhein, emodin, apigenin, kaempferol, protocatechuic acid, chlorogenic acid, caffeic acid, ferulic acid, catechin, epicatechin, ursolic acid, oleanolic acid and curcumin were purchased from Sigma Aldrich Ltd. (St. Louis, MO, USA). Analytical standards of (purity≥95%) quercetin, naringin, rutin and luteolin were purchased from Extrasyntheses (Genay, France). DMEM (Dulbecco's Modified Eagle's Medium) and FBS (Foetal Bovine Serum) were purchased from Sigma Aldrich Ltd. (St. Louis, MO, USA).

2.2 EXTRACTION AND SAMPLE PREPARATION

The dry plant parts of selected *Cassia* species were powdered to a homogeneous size by a pulverizer and sieved through a 40-mesh sieve, respectively. The dried powder of each part (5 g) was weighed precisely and sonicated with 200 mL of 100% methanol for 30 min at room temperature using an ultrasonic water bath (53 KHz) and left for 24 hours at room temperature. Three repeats of the extraction process were carried out on each individual sample. The solution was filtered through Whatman filter paper and evaporated to dryness under reduced pressure using rotatory evaporator

(Buchi Rotavapor-R2, Flawil, Switzerland) at 40°C. Dried residues (5 mg) were used for *in-vitro* anti-proliferative activity screening and (1 mg) were weighed accurately and dissolved in 1 mL of methanol using ultrasonicator (Bandelin SONOREX, Berlin) for UPLC-MS/MS analysis. The solutions were filtered through 0.22 µm syringe filter (Millex-GV, PVDF, Merck Millipore and Darmstadt, Germany). The filtrates were diluted with methanol to final working concentration. 1 µL of 1mg/ml sample was injected in HPLC-MS system for qualitative analysis. Internal standard (IS) (50 µL) was spiked in final working solution of quantitative experiments, vortexed for 30s and 5 µL aliquot was injected into the UPLC-MS/MS system for quantitative analysis.

2.3 PREPARATION OF STANDARD SOLUTION

A mixed standard stock solution containing anthraquinones (chrysophanic acid, physcion, rhein and emodin), phenolics (caffeic acid, ferulic acid, protocatechuic acid, chlorogenic acid), flavonoids (quercetin, kaempferol, apigenin, naringin, catechin, rutin, epicatechin and luteolin), triterpenoid (ursolic acid, oleanolic acid) was prepared in methanol. Mixed standards were diluted with methanol within the ranges from 0.5 to 250 ng/mL to prepare working standard solutions which was used for plotting calibration curve. Curcumin was used as the IS and MS parameter for curcumin were analysed in negative ESI mode. The 50 ppb concentration of IS curcumin was spiked to standards mixture of analyte and mixed properly. The standard stock and working solutions were all stored at –20 °C until use and vortexed prior to injection.

2.4 HPLC-ESI-QTOF-MS CONDITIONS

2.4.1 Instrumentation

The analyses were performed on an Agilent 1200 HPLC system consisted of a quaternary pump (G1311A), online vacuum degasser (G1322A), autosampler (G1329A), thermostatted column compartment (G1316C) and diode-array detector (G1315D).

2.4.2 HPLC Condition

The HPLC separation was carried out on Agilent Poroshell 120 EC-C18 column (50 mm × 4.6 mm, 2.7 µ). The mobile phase consisted of 0.1% formic acid aqueous solution (A) and acetonitrile (B) with flow rate of 0.4ml/min under the gradient program of 20 to 30% (B) for initial 5 min, then 30 to 40% (B) from 5 to 10 min, 40% (B) from 10 to 25 min, 40-50% (B) from 25 to 30 min, 20% (B) from 30 to 35 min. The sample injection volume was 1 µL.

2.4.3 Mass Spectrometric Conditions

Mass spectrometric analysis was performed on Agilent 6520 QTOF mass spectrometer in positive ESI mode. The resolving power of QTOF analyser was set above 15,000 (FWHM, full width at half maximum) and spectra were acquired within a mass range of m/z 100–1000. Nitrogen was used as nebulising, drying and collision gas. Capillary temperature was set to 350 °C and nebuliser pressure to 40 psi and the drying gas flow rate was 10 L/min. Ion source parameters such as Vcap, fragmentor, skimmer and octapole radio frequency peak voltage were set to 3500 V, 150 V, 65 V and 750 V, respectively. The chromatographic and mass spectrometric analyses, including the prediction of chemical formula and exact mass calculation, were performed by using Mass Hunter software version B.04.00 build 4.0.479.0 (Agilent Technology).

2.5 UPLC- ESI-QQQ_{LIT}-MS/MS CONDITIONS

2.5.1 Instrumental Parameters

The system used was an Acquity ultra-performance liquid chromatography (UPLC) consisting of an autosampler and a binary pump (Waters, Milford, MA) equipped with a 10 µL loop (partial loop injection mode). The UPLC system was coupled to triple-quadrupole linear ion trap mass spectrometer (API 4000 QTRAP™ MS/MS system from AB Sciex, Concord, ON, Canada) equipped with electrospray (Turbo V™) ion source was operated in negative ionization mode.

2.5.2 UPLC Condition

Column	Waters Acquity BEH C18 (2.1 mm × 50 mm, 1.7 µm)
Eluents	0.1% formic acid in water (A) and acetonitrile (B)
Gradient program	Initial linear increase from 5% to 90% B over 5 min, followed by hold of 90% B over 1 min, then return to the initial condition over 2 min
Flow rate	0.4 mL/min
Column thermostat	30 °C
Injection volume	5 µL

2.5.3 Mass Spectrometric Conditions

Scan type	Multiple-reaction monitoring (MRM)
Ion source	ESI (negative mode)
Collision-activated dissociation (CAD) gas	Medium
Interface heater	On
Polarity	Negative

2.5.4 Source Parameters Optimized for Negative Mode

Ion Spray voltage (IS)	–4200 V
Turbo spray temperature (TEM)	550 °C
Nebulizer gas (GS 1)	20 psi
Heater gas (GS 2)	20 psi
Curtain gas (CUR)	20 psi

2.5.5 Compound-Dependent Parameters

Optimization of the mass spectrometric conditions was carried out by infusing 50 ng/ml solutions of the analytes dissolved in methanol at 10 µl/min using a Harvard "22" syringe pump (Harvard Apparatus, South

Natick, MA, USA). For the MRM quantitation, highest abundance of precursor-to-product ions for each compound was chosen. The dwell time for both the parent and the IS was set at 200 ms. For, full scan ESI-MS analysis, the spectra covered the range from m/z 100 to 1000. Analyst 1.5.1 software package (AB Sciex) was used for instrument control and data acquisition.

2.6 PRINCIPAL COMPONENT ANALYSIS

PCA was carried out based on the contents of eighteen bioactive compounds in seed and leaf of five *Cassia* species, using STATISTICA 7.0 software. When the amount of investigated compounds were found below the quantitation limit or not detected than these values were considered to be zero.

2.7 OPTIMIZATION OF UPLC CONDITIONS

Complete separation of proximate analytes is certainly not required for MS/MS detection. In this study, chrysophanic acid and emodin are having same product ion, while catechin and epicatechin are having same precursor and product ion. To obtain better resolution various compositions of solvents were tried to get a suitable mobile phase. Due to stronger elution ability of acetonitrile over methanol, it was selected for this method. Similarly, an Acquity UPLC BEH C18 (2.1 mm × 50 mm, 1.7μm; Waters, Milford, MA) column which was more suitable for acidic mobile phase with smoother baseline was selected as compared to other tested columns. Formic acid was found more effective for ionization of compounds detected in the negative ESI mode. After testing various concentrations (0.05%, 0.1%, 0.2% and 0.3%) of formic acid 0.1% formic acid concentration was finally selected. A gradient elution with 0.1% formic acid in water and acetonitrile at a flow rate of 0.4 mL/min with the column temperature of 30 °C resulted in separation of the 18 compounds in less than 8 min chromatographic run time.

2.8 OPTIMIZATION OF MS/MS CONDITIONS

All the MS parameters for 18 compounds i.e., precursor ion, product ion, declustering potential (DP), entrance potential (EP), collision energy (CE) and cell exit potential (CXP) were optimized in negative ESI mode, by flow injection analysis (FIA). The chemical structures of 18 components were characterized based on their retention behaviour and MS information such as quasimolecular ions $[M-H]^-$, fragment ions $[M-H-COO]^-$, $[M-H-COO-CH_3]^-$, $[M- CO-H_2O]$ compared to related standards and literatures (Yu et al., 2009; Xia et al., 2011; Wei et al., 2013; Pandey et al., 2014). MRM parameters were optimized to achieve the most abundant, specific and stable MRM transition for each compound as shown in Table 2.1. MRM extracted ion chromatogram of 18 analytes are shown in Figure 2.1.

2.9 ANALYTICAL METHOD VALIDATION

The proposed UPLC-MRM method for quantitative analysis was validated according to the guidelines of international conference on harmonization (Guideline, I.H.T., Q2B (R1) 2005) by linearity, LOQs and LODs, precision, solution stability and recovery.

2.9.1 Linearity, LOD and LOQ

The stock solution was diluted with methanol to different working concentrations for the construction of calibration curves. The linearity of calibration was performed by the analytes-to-IS peak area ratios versus the nominal concentration and the calibration curves were constructed with a weight ($1/x^2$) factor by least-squares linear regression. The LODs and LOQs were measured with S/N of 3 and 10, respectively as criteria. The results were listed in Table 2.2.

All the calibration curves indicated good linearity with correlation coefficients (r^2) from 0.9986 to 0.9999 within the test ranges. The LODs for

TABLE 2.1 MRM parameters, retention time (Rt), declustering potential (DP), entrance potential (EP), collision energy (CE) and cell exit potential (CXP) for each analyte. (Reproduced from Chandra et al., 2015 with permission from Elsevier)

S. No.	R_t *(min)*	*Analyte*	*Q1 Mass (Da)*	*Q3 Mass (Da)*	*DP (V)*	*EP (V)*	*CE (eV)*	*CXP (V)*
1	1.05	Protocatechuic acid	153.0	109.0	−64	−5	−22	−9
2	1.15	Chlorogenic acid	353.0	191.0	−60	−10	−30	−10
3	1.29	Epicatechin	289.4	203.2	−120	−9	−22	−6
4	1.31	Caffeic acid	179.0	135.0	−48	−8	−21	−11
5	1.32	Catechin	289.1	203.0	−110	−10	−22	−10
6	1.64	Rutin	609.0	301.0	−197	−10	−45	−17
7	1.72	Ferulic acid	193.0	134.0	−58	−5	−23	−9
8	1.82	Naringin	579.0	271.0	−134	−10	−49	−6
9	2.32	Quercetin	301.0	151.0	−107	−9	−31	−12
10	2.32	Luteolin	285.0	133.0	−139	−10	−38	−12
11	2.66	Apigenin	269.0	117.0	−71	−5	−45	−9
12	2.67	Kaempferol	285.0	239.0	−95	−5	−39	−15
13	3.26	Rhein	283.1	182.8	−57	−8	−42	−10
14	3.85	Emodin	268.9	224.9	−94	−4	−34	−20
15	4.31	Chrysophanic acid	252.9	224.9	−115	−10	−39	−12
16	4.54	Physcion	282.9	239.8	−93	−7	−40	−20
17	5.21	Oleanolic acid	455.4	455.4	−120	−8	−7	−12
18	5.23	Ursolic acid	455.1	455.0	−130	−10	−9	−13

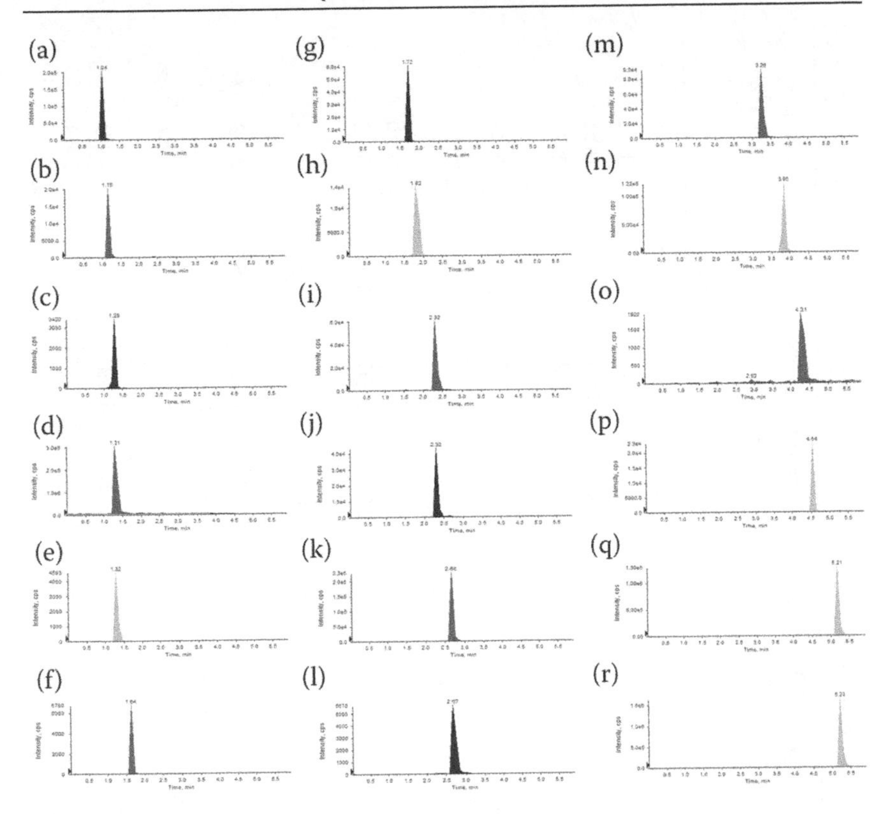

FIGURE 2.1 UPLC-MRM extracted ion chromatogram of analytes (a) protocatechuic acid, (b) chlorogenic acid, (c) epicatechin, (d) caffeic acid, (e) catechin, (f) rutin, (g) ferulic acid, (h) naringin, (i) quercetin, (j) luteolin, (k) apigenin, (l) kaempferol, (m) rhein, (n) emodin, (o) chrysophanic acid, (p) physcion, (q) oleanolic, (r) ursolic acid and IS i.e., (s) curcumin. (Reproduced from Chandra et al., 2015 with permission from Elsevier)

each analyte varied from 0.02–1.34 ng/mL and LOQs from 0.06–3.88 ng/ml and were much lower than those obtained with previous HPLC methods (Prakash et al., 2007; Ni et al., 2009; Chewchinda et al., 2012; Chewchinda et al., 2013; Chewchinda et al., 2014).

2.9.2 Precision, Stability and Recovery

Precision was determined by relative standard deviation (RSD) with intra-day and inter-day variations, for the determination of precision of the developed

TABLE 2.2 Validation parameters for 18 reference analytes. (Reproduced from Chandra et al., 2015 with permission from Elsevier)

Analytes	*Regression Equation*	r^2	*Linear Range ng/ml*	*LOD ng/ml*	*LOQ ng/ml*	*Precision RSD (%)*		*Stability*	*Recovery RSD (%)*
						Intra-day (n = 6)	*Inter-day (n = 6)*	*RSD (n = 5)*	
Protocatechuic acid	$y = 8.32x - 0.04$	0.9993	1–50	0.12	0.36	1.66	1.05	2.78	1.22
Chlorogenic acid	$y = 18.2x + 0.01$	0.9999	1–50	0.16	0.48	0.86	2.63	1.66	1.56
Epicatechin	$y = 2.69x - 0.02$	0.9991	1–100	0.19	0.56	2.22	2.94	1.68	0.55
Caffeic acid	$y = 12.68x + 0.09$	0.9995	1–50	0.08	0.28	1.72	1.82	2.11	1.27
Catechin	$y = 37.80x + 0.11$	0.9998	10–250	2.12	2.40	1.24	1.36	2.15	2.42
Rutin	$y = 3.64x - 0.01$	0.9995	1–50	0.18	0.52	1.13	1.14	3.19	1.66
Ferulic acid	$y = 9.75x - 0.01$	0.9991	5–100	1.10	1.43	2.51	1.23	2.86	0.86
Naringin	$y = 0.73x - 0.001$	0.9993	10–250	1.34	3.88	0.63	2.36	1.89	0.78
Quercetin	$y = 13.43x - 0.05$	0.9992	1–50	0.18	0.52	1.02	1.17	1.66	1.45
Luteolin	$y = 0.02x -0.01$	0.9998	0.5–50	0.06	0.31	0.77	2.25	1.59	2.38
Apigenin	$y = 1.13x + 0.22$	0.9998	0.5–100	0.04	0.29	3.37	0.93	2.68	1.89
Kaempferol	$y = 0.01x -0.003$	0.9999	0.5–100	0.02	0.27	0.88	0.91	2.45	1.87
Rhein	$y = 0.62x -0.09$	0.9986	5–100	0.66	1.95	1.52	4.09	2.16	1.59
Emodin	$y = 34.00x - 1.3$	0.9997	1–100	0.20	0.55	0.62	1.77	0.39	0.63

Chrysophanic acid	$y = 28.81x + 0.6$	0.9990	1–100	0.15	0.49	1.58	2.10	1.20	2.11
Physcion	$y = 252.91x + 3518.2$	0.9995	10–1000	2.12	2.38	0.62	1.36	1.29	1.92
Oleanolic acid	$y = 1.07x + 2.2$	0.9997	1–100	0.02	0.57	0.38	1.38	0.64	0.58
Ursolic acid	$y = 0.03x + 5.65$	0.9998	0.5–100	0.04	0.22	0.44	1.27	0.72	1.36

method were evaluated by determination of 18 analytes in six replicates on a single day and by duplicating the experiments over three successive days. The overall intra-day and inter-day precision was not more than 3.37%. Stability of sample solutions stored at room temperature was evaluated by replicate injections at 0, 2, 4, 8, 12 and 24 h. The stability RSD% value of eighteen analytes is ≤ 3.19%.

To evaluate the accuracy, recovery test was applied by spiking three different concentration levels (high, middle and low) of the analytical standards into the samples. Three replicates were performed at each level. The analytical method developed had good accuracy with overall recovery in the range from 97.75% to 105.09% (RSD ≤ 2.42%) for all analytes (Table 2.2).

2.10 QUALITATIVE ANALYSIS OF *CASSIA OCCIDENTALIS* SAMPLES

Metabolic profiling of ethanolic extract of *Cassia* occidentalis sample resulted in detection of 24 compounds. All the compounds were tentatively identified are reported in Table 2.3. The identification of each compound was tentatively confirmed on the basis of their retention time, accurate mass and molecular formula by HPLC-ESI–QTOF–MS. The base peak chromatogram (BPC) of ethanolic extract of *Cassia* occidentalis is given in Figure 2.2.

2.11 QUANTITATIVE ANALYSIS OF *CASSIA* SPECIES SAMPLES

The proposed UPLC-ESI-MS/MS method was applied to quantify 18 bioactive compounds in different plant parts of five *Cassia* species viz., *C. auriculata, C. fistula, C. occidentalis*, *C. siamea* and *C. uniflora.* Remarkable differences in the quantity of anthraquinones, phenolics, flavonoids and terpenoids were observed during the study of quantitative result. In respect to the content of each analyte, chrysophanic acid, physcion, ursolic acid, oleanolic acid, rutin and luteolin are the most abundant bioactive compounds in majority of parts of *Cassia* species as listed in Table 2.4.

Anthraquinone like rhein (14600 μg/g) was present in highest amount in pulp of *C. fistula,* while found below detection limit in other species, which

TABLE 2.3 List of identified phytochemicals in ethanolic extract of *Cassia* by HPLC-ESI–QTOF–MS

S. No.	*RT (min)*	*Cal. Mass [M-H]⁻*	*Obs. Mass [M-H]⁻*	*Error (Δppm)*	*Molecular Formula*	*Identification*
1	11.9	593.1512	593.1514	–0.16	$C_{27}H_{30}O_{15}$	Kaempferol 3-O-rutinoside
2	12.3	447.0933	447.093	0.56	$C_{21}H_{20}O_{11}$	Kaempferol 3-O-glucoside
3	12.4	447.0933	447.0931	0.37	$C_{21}H_{20}O_{11}$	6-Hydroxyluteolin-5-rhamnoside
4	14.5	431.0984	431.0985	0.25	$C_{21}H_{20}O_{10}$	Vitexin
5	14.6	431.0984	431.0985	0.21	$C_{21}H_{20}O_{10}$	Isovitexin
6	15.0	167.035	167.0351	0.62	$C_8H_8O_4$	Vanillic Acid
7	16.0	607.1668	607.1668	0.05	$C_{28}H_{32}O_{15}$	Physcion 8-gentiobioside
8	17.6	253.0506	253.0507	–0.33	$C_{15}H_{10}O_4$	Chrysophanol
9	17.7	577.1563	577.1561	0.32	$C_{27}H_{30}O_{14}$	Kaempferide-3-rhamnoside7xyloside
10	18.0	269.0455	269.0457	0.7	$C_{15}H_{10}O_5$	3',4',7-trihydroxyflavone
11	18.1	269.0455	269.0454	0.5	$C_{15}H_{10}O_5$	Apigenin
12	18.2	269.0455	269.0453	1.02	$C_{15}H_{10}O_5$	Emodin
13	18.5	153.0196	153.0196	–1.65	$C_7H_6O_4$	Protocatechuic acid
14	18.9	313.0718	313.0708	2.95	$C_{17}H_{14}O_6$	4'-Methylcapillarisin
15	19.3	285.0405	285.0394	3.56	$C_{15}H_{10}O_6$	Luteolin
16	20.5	253.0506	253.0508	–0.74	$C_{15}H_{10}O_4$	4',7-dihydroxyflavone
17	20.6	253.0506	253.0501	2.18	$C_{15}H_{10}O_4$	Chrysophanic acid
18	20.9	283.0612	283.0612	–0.05	$C_{16}H_{12}O_5$	Physcion
19	21.2	287.0561	287.0561	0.05	$C_{15}H_{12}O_6$	Eriodictyol
20	21.4	243.0663	243.0659	1.46	$C_{14}H_{12}O_4$	Piceatannol
21	22.7	285.0405	285.0403	0.43	$C_{15}H_{10}O_6$	Kaempferol

(Continued)

TABLE 2.3 (*Continued*)

S. No.	*RT (min)*	*Cal. Mass [M-H]⁻*	*Obs. Mass [M-H]⁻*	*Error (Δppm)*	*Molecular Formula*	*Identification*
22	23.3	301.0354	301.0351	0.91	$C_{15}H_{10}O_7$	Quercetin
23	26.4	271.0612	271.0615	–0.94	$C_{15}H_{12}O_5$	Naringenin
24	27.0	271.0612	271.0612	0.08	$C_{15}H_{12}O_5$	Isotoralactone

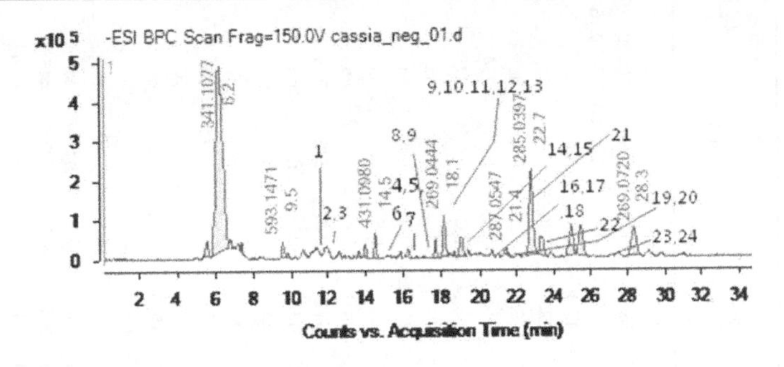

FIGURE 2.2 Base peak chromatogram of ethanolic extract of *Cassia occidentalis.*

clearly showed the characteristic variation among five *Cassia* species. Other anthraquinones like emodin (2755 μg/g), chrysophanic acid (136600 μg/g) and physcion (5950 μg/g) considered as the characteristic components were detected highest in *C. siamea* root. In the similar pattern for terpenoids, oleanolic acid (10560 μg/g) and ursolic acid (39450.16 μg/g) was detected highest in the root of *C. siamea* while found comparatively low and/or not detected in other species, which shows a characteristic pattern and significant variation among five *Cassia* species. This clearly indicated that *C. siamea* root may be used as a substitute for *C. auriculata* and *C. fistula* in various herbal formulations/preparations.

Higher content of flavonoids such as rutin (45550 μg/g) and quercetin (5440.23 μg/g) were detected in the leaves and luteolin (12200.0 μg/g) in flowers of *C. uniflora* with comparative lower content in *C. occidentalis* and *C. fistula.* But in contrast, kaempferol (3280 μg/g) was detected highest in pulp of *C. fistula*. Phenolics such as protocatechuic acid (4215 μg/g) and ferulic acid (236 μg/g) were also found highest in flower and seed of *C. uniflora* respectively, while not detected in other parts of *Cassia* species. Epicatechin (12350.32 μg/g) was detected in significant highest amount in stem of *C. fistula*, which is in agreement to the reported literature (Patidar et al., 2012). Graphical representations of these observations were shown in Figure 2.3, which clearly explains the chemical variations among five *Cassia* species and their plant parts.

The graphical representation of 18 compounds content in different plant parts of *Cassia* species shown in Figure 2.4. Contents of the bioactive compounds are directly responsible for the efficacy hence change in their concentration may lead to the variation. Therefore, it is important to select as many as possible pharmacologically active chemical markers in order to distinguish five *Cassia* species with plant parts of different quantities. The

TABLE 2.4 Contents of the 18 reference analytes in *Cassia* species (μg/g). (Reproduced from Chandra et al., 2015 with permission from Elsevier)

Samples	*Protocatechuic Acid*	*Chlorogenic Acid*	*Epicatechin*	*Caffeic Acid*	*Catechin*	*Rutin*	*Ferulic Acid*	*Naringin*	*Quercetin*	*Luteolin*	*Apigenin*	*Kaempferol*	*Rhein*	*Emodin*	*Chrysophanic acid*	*Physcion*	*Oleanolic Acid*	*Ursolic Acid*
C. siamea_LF	13.4	3035.2	24.4	7350	bdl	625	7.50	nd	bdl	1830	1520	nd	1360.43	117	137.4	880.11	110.25	66.0
C. siamea_ST	33.62	4835	nd	555	nd	645	55.5	nd	246	2115.41	nd	nd	nd	195.23	592.12	2925.22	3500	10300
C. siamea_RT	1125	125.8	232	1670.26	20.61	830	37.35	nd	3140	1440.45	nd	827	nd	2755	13660	5950	105600	39450.16
C. siamea_SD	bdl	213.16	nd	156.5	2.66	250.13	nd	nd	nd	925	bdl	nd	nd	84.5	17.40	bdl	nd	nd
C. siamea_FL	4.88	1495	nd	477.5	10.28	244	7.75	nd	1.46	2050	nd	961	nd	121	22.6	1850.55	nd	17.78
C. fistula_LF	22.1	180	nd	322.5	1356.23	2205	10.44	61.20	32.64	950	nd	nd	995.2	87.51	79.15	-	nd	nd
C. fistula_ST	307.6	248	12350.32	1200	1662.65	5350	127.16	nd	67.8	1785	bdl	nd	7890	156.5	114.8	179.5	69.4	334
C. fistula_RT	41.7	193	nd	149	1785.13	690.16	nd	nd	bdl	1010	bdl	nd	866	98.5	164.4	32.0	14.74	37.2
C. fistula_SD	38.0	178.46	nd	52.5	bdl	665.24	nd	nd	nd	935	bdl	nd	bdl	93.0	nd	bdl	nd	nd
C. fistula_PL	32.1	187	2360	145	1668	895	nd	nd	nd	930	nd	3280	14600	84.13	245.33	43.35	bdl	nd
C. occidentalis_LF	7.00	189.52	nd	223.5	bdl	590.34	63.5	120.81	nd	1100	nd	nd	nd	93.52	bdl	58.5	nd	nd
C. occidentalis_ST	bdl	191.5	nd	289.5	16.6	262	86.14	1.56	nd	1325	71.7	nd	nd	145	86.0	95.6	nd	nd
C. occidentalis_RT	114	197.5	bdl	177.3	bdl	250	14.55	nd	65.0	4385	334	nd	bdl	560	612	1030	310	776
C. occidentalis_SD	nd	197.5	nd	149.5	24.3	610.24	bdl	nd	bdl	1245	nd	nd	bdl	126.5	nd	303.5	nd	nd
C. occidentalis_FL	nd	187	nd	110	bdl	bdl	bdl	nd	bdl	935	bdl	66.2	bdl	84.5	nd	nd	nd	nd
C. auriculata_LF	nd	220.23	54.2	156.5	25.6	16650.12	6.66	nd	402	1160.35	bdl	nd	bdl	89.5	nd	nd	nd	nd
C. auriculata_ST	nd	197.5	26400	204.5	2440	5550	36.4.25	bdl	354	1095	nd	nd	bdl	106	67.6	35.6	119.4	169.2
C. auriculata_RT	36.4	180.5	1638	130	1002	635	3.06	bdl	57.0	930	nd	bdl	nd	164.5	484	276	478	1044
C. auriculata_SD	nd	198.5	36.4	174.5	30.8	4685.11	42.0	bdl	11.48	1050	bdl	bdl	nd	233	15.74	19.65	nd	nd
C. auriculata_FL	bdl	88.2	119.2	305.5	66.2	9000	nd	bdl	508	2005	bdl	bdl	nd	98.0	nd	nd	nd	nd
C. uniflora_LF	1215	213.5	398	92.5	332	45550	55.4	nd	5440.23	9050	bdl	bdl	bdl	122.5	nd	35.4	7.08	24.2
C. uniflora_ST	570	76.8	nd	201.5	1.13	650	92.8	nd	nd	1040	bdl	nd	bdl	132.0	21.4	43.1	nd	nd
C. uniflora_RT	159	183.5	21.75	194.5	34.2	625.23	172.8	nd	9.24	1215.3	nd	nd	nd	213.5	950	267.5	172.6	446.31
C. uniflora_SD	4020	86.6	35.4	645	60.6	8000.25	236	nd	1706	9250	nd	nd	nd	745.0	nd	nd	nd	52.5
C. uniflora_FL	4215	83.2	151.2	112.5	104	9950	137.2	nd	3640	12200	nd	nd	nd	125.5	34.1	2.52	24.2	6.32

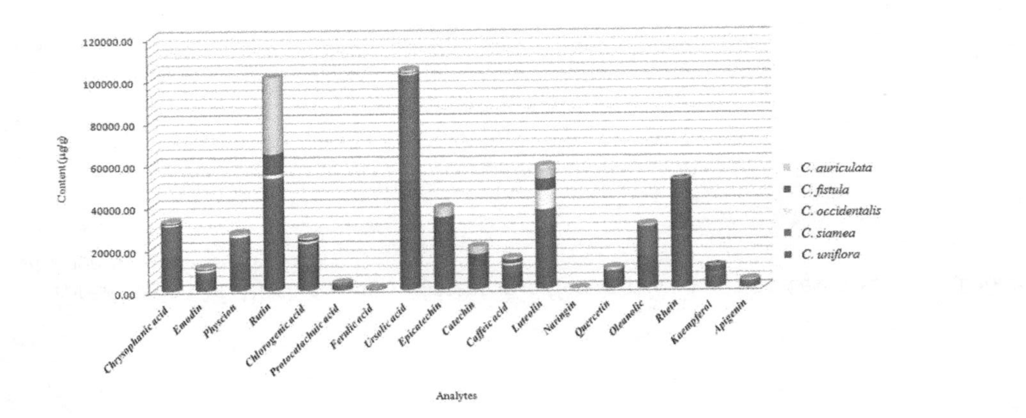

FIGURE 2.3 Graphical representation of content (µg/g) of 18 compounds in five *Cassia* species. (Reproduced from Chandra et al., 2015 with permission from Elsevier)

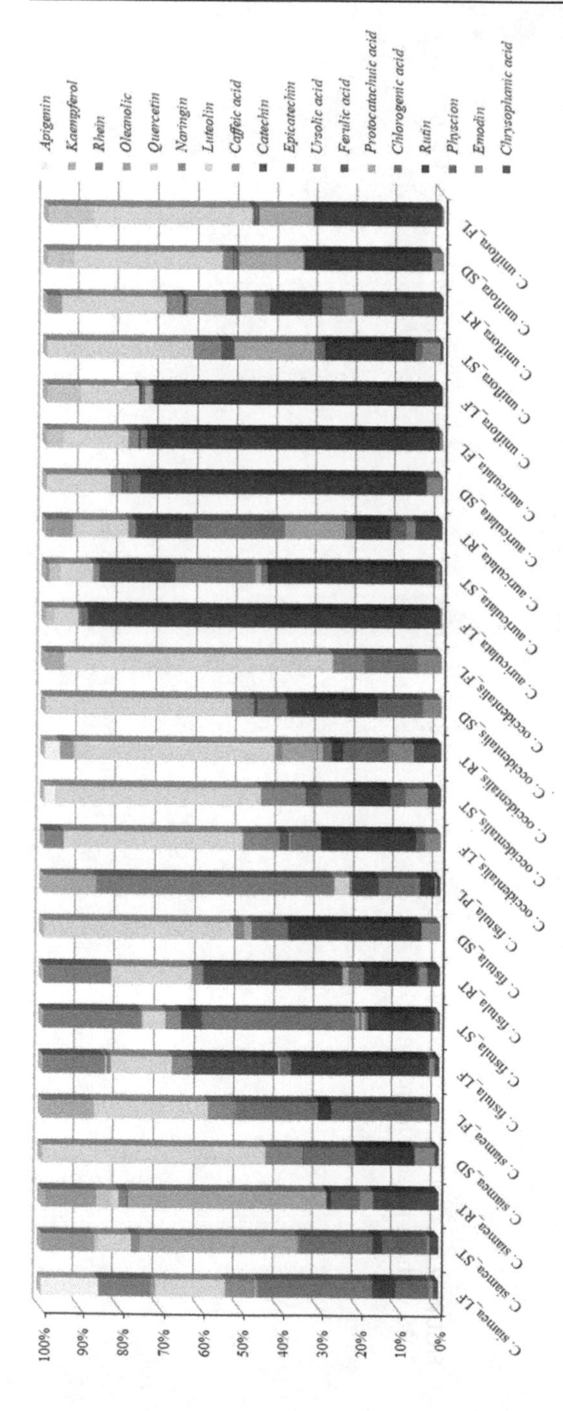

FIGURE 2.4 Graphical representation of 18 compounds content in different plant parts of *Cassia* species. (Reproduced from Chandra et al., 2015 with permission from Elsevier)

quantitative results also indicated the importance of plant parts and their impact on product quality. The overall quantitative analysis indicated that this method has significant importance in a comprehensive evaluation of selected 18 compounds, which could be used as the markers for quality control of *Cassia* species and its related preparations.

2.12 PRINCIPAL COMPONENT ANALYSIS TO STUDY CHEMICAL VARIATIONS

PCA is the most widely used chemometrics method for multivariate data analysis (Senousy et al., 2014). PCA was carried out based on the quantitative data to compare and evaluate the quality of species based on the characteristics of the contents of 18 investigated compounds in leaves, stem and root of five *Cassia* species.

The PCA was studied on the basis of total matrix data of 18 compounds collectively in all plant parts of five species. The PCA shows that the data matrix reduced to 5 PCs explaining 86% variation. Jointly the first two PCs explain 54.76% variation of data where the 18 compounds were distributed in 3 clusters in the loading biplot as shown in Figure 2.5A. This explained similarity in the pattern of quantities among plant parts of various species. However, score plot by PC1 vs PC2 as shown in Figure 2.5B indicated distribution of investigated samples in four clusters. From the loading scatter plot, it was observed that different variables show different contributions in discrimination of *Cassia* species. However quantitative analysis showed that leaf, stem and root of *C. siamea,* with leaf and seed of *C. uniflora* species were represented the characteristic pattern. The *C. siamea* root is placed in an extreme of the biplot indicating high quantity levels of most of the compounds.

PCA was also performed on the basis of whole data in individual plants parts viz., leaf, stem, root, flower, seed/pulp of five *Cassia* species and explained in Figure 2.6(a)–(e), respectively. PCA plot from leaf of *Cassia* species was able to explain 84.19% variation in data with PC1 and PC2 as shown in Figure 2.6(a). *C. uniflora* showed highest dominance due to similarity in bioactive compounds that is rutin, protocatechuic acid, epicatechin and quercetin. Similarly, PCA plots from stem of *Cassia* species explained 69.75% variation in data by the first two PCs shown in Figure 2.6(b). The first PC was dominated by the highest quantity of chrysophanic acid, emodin, physcion, chlorogenic acid, ursolic acid and oleanolic acid in *C. siamea. C.*

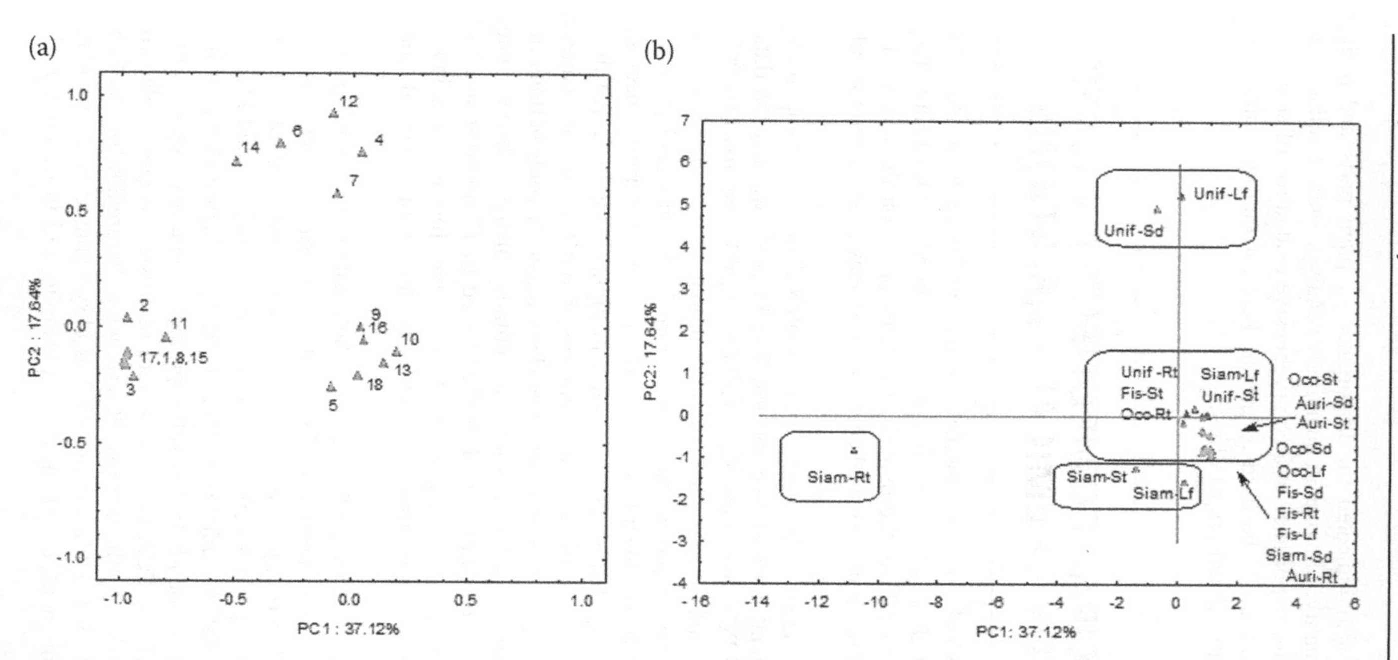

FIGURE 2.5 PCA loading plot of 18 compounds distributed in different *Cassia* species and PCA score plot of different plants parts (leaves, stem, root, flower, seed/pulp) of *Cassia* species. (Reproduced from Chandra et al., 2015 with permission from Elsevier)

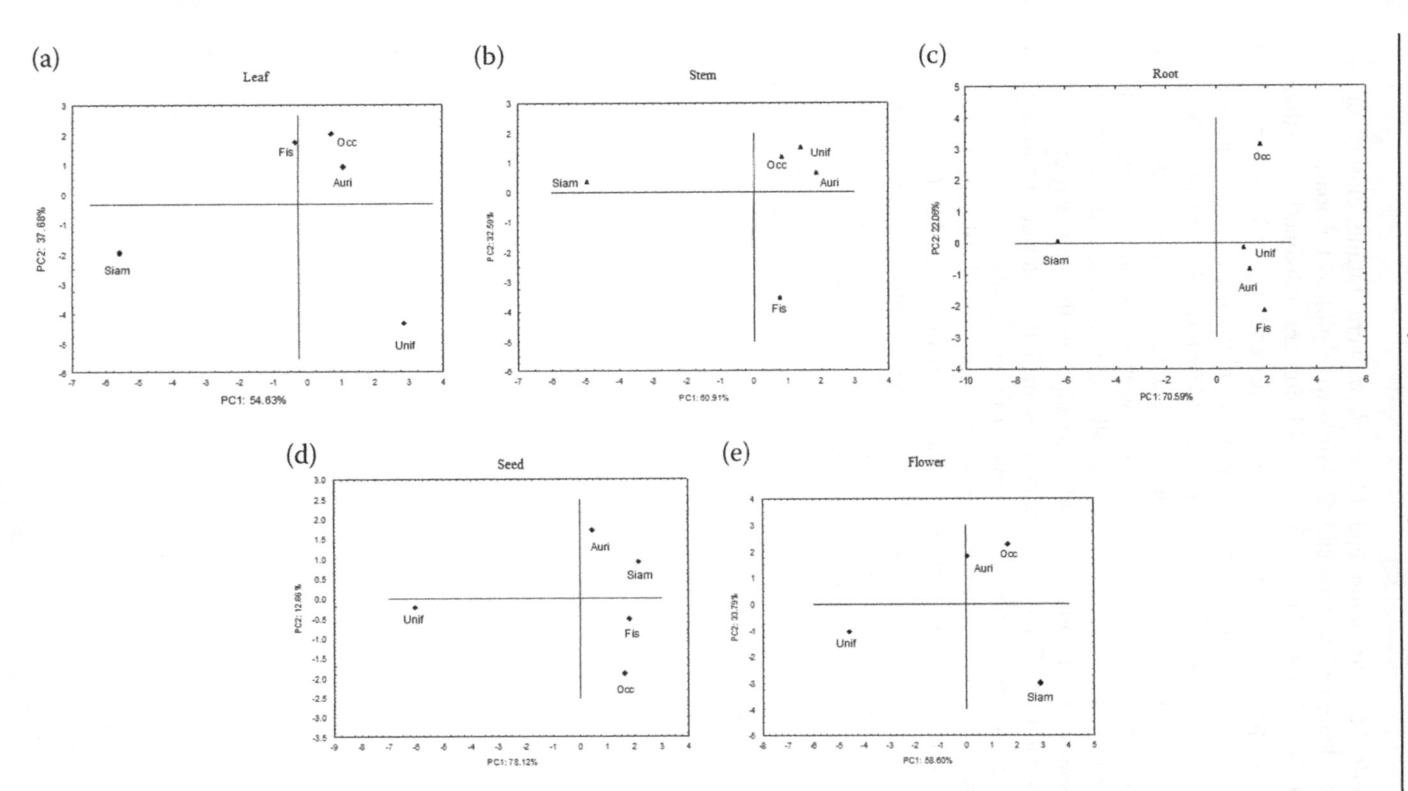

FIGURE 2.6 (a–e): PCA score plot of various plants parts of *Cassia* species: (a) leaf, (b) stem, (c) root, (d) seed, (e) flower. (Reproduced from Chandra et al., 2015 with permission from Elsevier)

occidentalis, *C. auriculata* and *C. uniflora* were falling closely in first quadrant whereas *C. siamea* and *C. fistula* were plotted far apart.

PC1 vs PC2 plot of roots shown in Figure 2.6(c) distinctly classified the five *Cassia* species. *C. siamea*, *C. occidentalis* and *C. fistula* were placed in the extremes of the two-dimensional plot due to the multidimensional pattern of the compounds. *C. uniflora* and *C. auriculata* have shown similarity in the pattern due to quantity of compounds. *C. siamea* root was identified the richest source of compounds due to chrysophanic acid, emodin, physcion, protocatechuic acid, ursolic acid, oleanolic acid and quercetin. PCA plot of *Cassia* species seed showed 90.78% variation by the first two components with PC1 and PC2. *C. uniflora* was lying on the left-hand side due to negative loadings of PC1 in the biplot as shown in Figure 2.6(d).

PCA plot of flowers explained 92.38% variation dominated by the negative loading effects in both PCs as shown in Figure 2.6(e). *C. auriculata* and *C. occidentalis* have fallen in the first quadrant due to some similarity in quantity of compounds. *C. siamea* and *C. uniflora* were in different quadrants. Depending upon the multidimensional pattern in the quantity of compounds, all *Cassia* species were individually classified and discriminated. Thus, the PCA study clearly showed the remarkable variations among the various plant parts of *Cassia* species.

In-vitro Anti-Proliferative Screening in Selected *Cassia* Species

3

Anti-proliferative activity is the ability of a compound to stop the growth of cells. It is used o treat cancer. Screening for *in-vitro* anti-proliferative activity of different *Cassia* species plant parts studied for their inhibitory effects on the growth of DLD 1 (colorectal adenocarcinoma), A549 (lung carcinoma), MCF-7 (breast adenocarcinoma), DU145 (prostate carcinoma) and FaDu (pharyngeal carcinoma) cell lines. Principal component analysis of analytical results depicted remarkable chemical differences and variations of bioactive compounds in different *Cassia* species. The present comprehensive evaluation is of great importance for the rational application and quality control of different *Cassia* species.

3.1 PLANT MATERIAL

Plant materials (leaf, stem, root, seed and flower of *Cassia auriculata, Cassia occidentalis, Cassia siamea, Cassia uniflora,* with fruit pulp instead of flower in case of *Cassia fistula*) were collected from the plants grown in western ghats of Maharashtra region between the months of May to September 2013. Voucher specimens of *C. auriculata* (ARC-5), *C. fistula* (ARC-2), *C. occidentalis* (ARC-3), *C. siamea* (ARC-1) and *C. uniflora* (ARC-4) were deposited in the Botanical Survey of India, Western Region Centre, Pune, India.

DOI: 10.1201/9781003186281-3

3.2 EXTRACTION AND SAMPLE PREPARATION

The dry plant parts of selected *Cassia* species were powdered to a homogeneous size by a pulverizer and sieved through a 40-mesh sieve, respectively. The dried powder of each part (5 g) was weighed precisely and sonicated with 200 mL of 100% methanol for 30 min at room temperature using an ultrasonic water bath (53 KHz) and left for 24 hours at room temperature. Three repeats of the extraction process were carried out on each individual sample. The solution was filtered through Whatman filter paper and evaporated to dryness under reduced pressure using a rotatory evaporator (Buchi Rotavapor-R2, Flawil, Switzerland) at 40°C. Dried residues (5mg) were used for *in-vitro* anti-proliferative activity screening.

3.3 CELL LINES AND ANTI-PROLIFERATIVE ASSAY

Human cancer cell lines A549 (lung carcinoma), MCF-7 (breast adenocarcinoma), DU 145 (prostate carcinoma), DLD1 (colorectal adenocarcinoma) and FaDu (squamous cell carcinoma of pharynx) were obtained from American Type Culture Collection (ATCC), USA. These cells were cultured in DMEM supplemented with 10% FBS and antibiotic combinations in 5% CO_2 humidified atmosphere at 37°C.

A colorimetric sulforhodamine B (SRB) assay is the most preferred technique and was used for the measurement of anti-proliferative activity (Fricker and Buckley, 1995; Keepers et al., 1991; Skehan et al., 1990; Adaramoye et al., 2011). This basically depends on the incur of the negatively charged pink amino xanthine dye, SRB through basic amino acids in the cells. The dye released will give more acute colour and more absorbance, when the number of cells and amount of dye is taken up is greater, after fixing, the cells are lysed (Skehan et al., 1990). The SRB assay is sensitive, reproducible and gives better linearity, a good signal-to-noise ratio and has a stable end-point that does not require a time-sensitive measurement (Fricker and Buckley, 1995;

Keepers et al., 1991). Ten thousand cells were seeded to each well of 96-well plate, grown overnight and exposed to test samples at 100 µg/ml concentration for 48 h. Cells were then fixed with ice-cold trichloroacetic acid (50% w/v, 50 µl/well), stained with SRB (0.4% w/v in 1% acetic acid, 50µl/well), washed and air dried. Bound dye was dissolved in 150 µL of 10 mM Tris base and plates were read at 510 nm absorbance (Epoch Microplate Reader, Biotek, USA).

Anti-proliferative activity of test samples was calculated as:

% inhibition in cell growth = [100-(Absorbance of compound treated cells/ Absorbance of untreated cells)] x100.

3.4 *IN-VITRO* ANTI-PROLIFERATIVE ACTIVITY EVALUATION

Evaluation of *in-vitro* anti-proliferative activity by SRB assay against the DLD1, A549, MCF-7, DU145 and FaDu cell line was done with doxorubicin, which is used as a positive control with percentage growth inhibition at 100 µg/ml concentration. Doxorubicine caused more than 70% growth inhibition in all the cell lines. Result in Table 3.1 showed that root of *C. siamea* (74.01%) and *C. auriculata* (91.96%) and leaf of *C. uniflora* (68.59% and 73.28%) were able to inhibit growth in DLD1 and FaDu cell lines, respectively. Similarly, leaf of *C. uniflora* (91.96%), stem of *C. auriculata* (90.62%) extracts showed >70% growth inhibitory effect against Du145 cell line. Seed and stem of *C. auriculata* (54.97% and 79.60%) were able to inhibit >50% growth in A549 and MCF-7 cell lines. The graphical representation of *in-vitro* anti-proliferative screening of different cell lines in different plant parts of *Cassia* species is shown in Figure 3.1.

TABLE 3.1 Percentage (%) growth inhibition in various plant parts of *Cassia* species. (Reproduced from Chandra et al., 2015 with permission from Elsevier)

	Percentage growth inhibition at 100μg/ml concentration				
	DLD1	*FaDu*	*DU145*	*A549*	*MCF-7*
C. auriculata -LF	49.99	75.1	77.22	18.23	59.54
C. auriculata -ST	65.08	63.59	90.62	46.55	79.6
C. auriculata -RT	68.59	73.28	91.96	44.9	51.6
C. auriculata -SD	51.92	68.82	40.17	54.97	56.04
C. auriculata -FL	67.61	63.48	67.57	33.45	39.02
C. fistula -LF	27.05	6.39	27.71	36.22	46.06
C. fistula -ST	26.51	10.53	37.76	14.82	8.86
C. fistula-RT	25.38	7.40	23.55	13.61	15.69
C. fistula -PL	14.89	−14.72	6.33	13.35	−10.72
C. fistula -SD	11.11	−25.33	7.34	−1.9	16.24
C. occidentalis -LF	33.48	22.34	25.09	23.65	27.63
C. occidentalis -ST	18.88	−0.97	−8.48	−5.43	−6.51
C. occidentalis -RT	13.98	23.2	6.20	10.14	3.62
C. occidentalis -SD	36.02	30.91	36.08	2.25	0.95
C. occidentalis -FL	12.76	24.44	28.78	9.64	31.54
C. siamea -LF	12.1	5.69	15.51	14.64	31.49
C. siamea ST	14.69	−7.32	−2.45	−1.16	2.41
C. siamea -RT	70.92	74.01	55.91	52.26	62.74
C. siamea -SD	7.10	−10.26	−3.79	1.57	−4.17
C. siamea FL	19.13	7.51	−31.99	−17.17	−16.44
C. uniflora -LF	68.59	73.28	91.96	44.90	51.6
C. uniflora -ST	15.48	36.33	20.60	16.44	15.01
C. uniflora RT	25.67	46.75	26.16	4.43	4.30
C. uniflora -SD	19.94	55.69	32.06	44.99	50.69
C. uniflora -FL	31.38	68.78	43.32	47.49	48.8
Doxorubicin	83.35	98.7	70.99	72.58	82.01

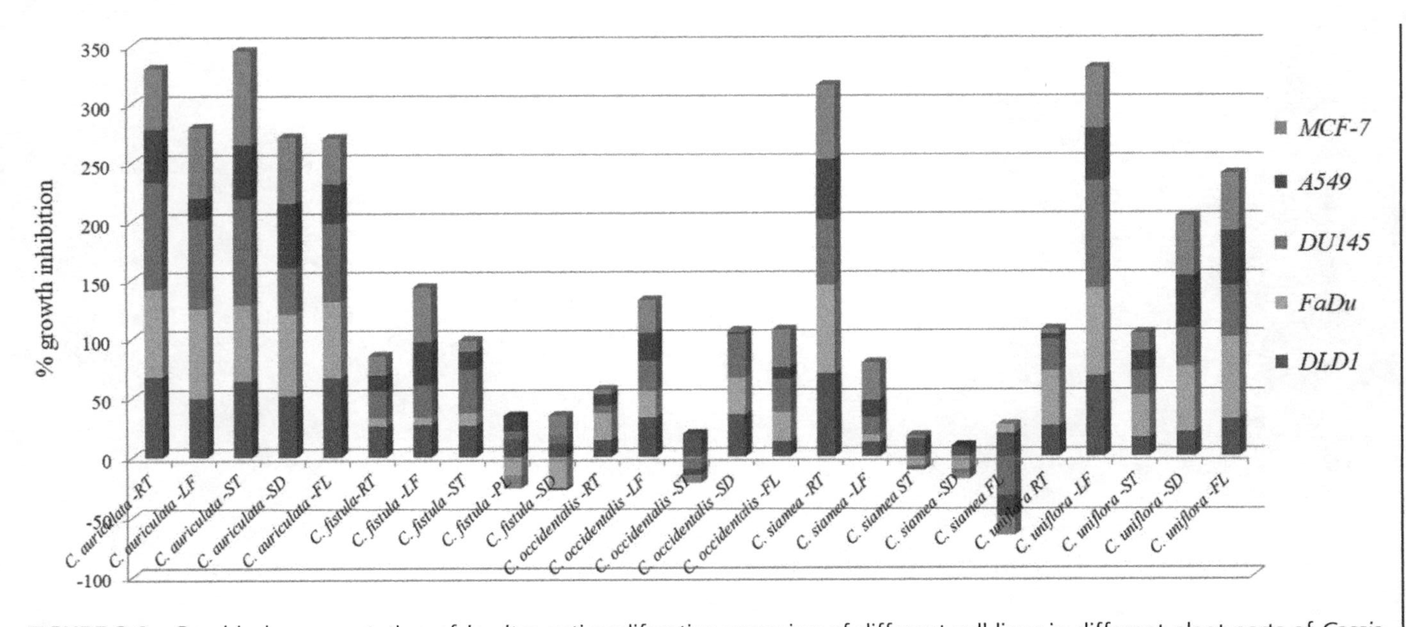

FIGURE 3.1 Graphical representation of *in-vitro* anti-proliferative screening of different cell lines in different plant parts of *Cassia* species. (Reproduced from Chandra et al., 2015 with permission from Elsevier)

Bone Regeneration Effect of *Cassia occidentalis* Linn. Extract and Its Isolated Compounds

4

4.1 BONE HEALING PROPERTY OF *CASSIA OCCIDENTALIS*

The topical application of *Cassia occidentalis* (CO) extract for bone healing is in practice for over a century in some areas of Andhra Pradesh, India. Inspired by this ethnotraditional practice, we investigated the phytochemistry of aerial parts of CO to make a standardized extract and fraction for oral application to treat specific disease conditions such as fracture healing and glucocorticoid (GC) hormone-induced osteoporosis (GIO). Developing in line with the recent phytopharmaceutical guideline of GoI, we have standardized the extract with as many as six osteogenic marker compounds (requirement is four) and identified five of those in plasma after oral dosing of the extract to rats. A hydroalcoholic extract of leaf and stem of CO for up to 5 g/kg was found to be well tolerated and safe in acute toxicity study and up to 2.5 mg/kg in 30-day

DOI: 10.1201/9781003186281-4

subacute (repeat dose) study in rats (Silva et al., 2011). This report however lacks data on cardiovascular, central nervous system and respiratory safety that are necessary prerequisites for human trial.

Plethora of phytochemicals isolated from various parts of CO mostly fall within flavonoids, anthraquinones and essential oils. In *Cassia* stem extract (CSE), ten of these were identified out of which six including apigenin, isovitexin, emodin, luteolin, 4′,7′-dihydroxyflavone and 3′,4′, 7-trihydroxyflavone showed osteogenic activity *in-vitro*. Besides promoting osteoblast function, these compounds and others present in CO (kaempferol, caryophyllenes and quercetin) have been reported to regulate the function of other bone cells including osteoclasts, chondrocytes and mesenchymal stem cells (MSCs). Consequently, compounds present in leaf and stem prevent osteoporosis, promote fracture healing and mitigate arthritis and joint degeneration. Thus, it is conceivable that CSE would protect against ovariectomy (OVX)-induced bone loss and RA. Action mechanisms include protection of osteoblasts and chondrocytes from oxidative damage, suppression of NFκB signaling and actions of pro-inflammatory cytokines, and regulation of a variety of MAPKs.

To establish the standardized extract of CO for the treatment of diseases of bone, IND-enabling studies prerequisite for taking up clinical trial should be carried out. Subsequently, human safety and tolerability in normal adults along with PK studies tracking blood levels of the osteogenic markers and concomitant identification of serum levels of bone formation marker (P1NP) would allow determination of safety of and PK/PD relationship for CO extract in phases 0–1 trial setting. After establishing human safety and tolerability, proof of concept (PoC) study in a limited number of GIO patients in a phase II trial to determine the efficacy dose and safety in patients would be reasonable given the efficacy of CO extract shown in preclinical model. Since high doses of GCs reduce serum P1NP, the bone formation marker and preclinical studies have shown an increase in serum P1NP in GC-treated rats by CSE; this bone biochemical marker could be monitored as the primary outcome measure of the treatment efficacy. Following a successful outcome of phase II trial, multicentric double-blind, placebo-control trials involving a large number of patients will be necessary to finally demonstrate the efficacy of CO extract in mitigating GIO in comparison to standards of care [a bisphosphonate drug (alendronate or risedronate; anti-resorptive drug) and teriparatide (injectable PTH, bone anabolic drug)]. Final efficacy measure will be decrease in fracture incidences by the CO extract equivalent to or better than the standards of care with better safety and tolerability profile.

4.2 OSTEOPOROSIS AND ARTHRITIS

The effects of CO extracts and the constituent compounds will be discussed concerning the healing of fracture and diseases of bone loss and joint degeneration. Fracture healing is a multi-stage process and the events of healing recapitulate that of bone development (Morgan et al., 2014). Chondrocytes and osteoblasts are crucial players in fracture healing, and their functions are impaired in aging and osteoporotic condition (Clark et al., 2017). Osteoporosis or loss of bone mineral density (BMD) is primarily caused due to aging that is associated with loss of gonadal function. In women, reduced estrogen levels (perimenopausal stage) and complete estrogen deficiency (menopausal stage) act as triggers of bone loss (osteopenia) which occur between the fourth and fifth decades of life (Clarke and Khosla, 2010). If osteopenia remains untreated, then it leads to osteoporosis that in turn significantly increases the risk of fractures. The two major sites of such fractures include spine and femur neck (which represents hip fracture) and are characteristically caused by non-traumatic falls (Sozen et al., 2017). In men, decreases in androgen levels commonly occurring from the fifth decade of life result in bone loss and the development of osteoporosis. Besides aging-related osteoporosis (so-called primary osteoporosis), many diseases including renal failure, liver impairment, Cushing's disease, multiple sclerosis, chronic obstructive pulmonary disease, anorexia nervosa, hyperparathyroidism, hyperthyroidism, hypercortisolism, etc (Mirza and Canalis, 2015). Several medications including corticosteroids, barbiturates, and aluminium-containing antacids besides others cause osteoporosis. The first line of osteoporosis therapy is bisphosphonates (BPs) that suppress the function of osteoclasts, the bone resorptive cells. Although these drugs suppress further bone loss, fail to restore the lost bones due to osteoporosis (Van Beek et al., 2002). BPs or for that matter denosumab, a more recent anti-resorptive drug that acts by neutralizing RANKL suppresses osteoblast function given that in the bone remodelling process, osteoblast function is coupled to osteoclast function (Hanley et al., 2012; Teitelbaum, 2016). Although BPs reduce the risk of any fracture yet associated with atypical thighbone fracture unprovoked by major trauma and is a debilitating condition (Kharwadkar et al., 2017). This association is stronger than the one between smoking and lung cancer. Several studies have reported that long-term use of BPs (>3 years) negatively impacts bone quality assessed by the biomaterial composition of bones. Because of its very high affinity to bone, BPs bind and stay in bones several years after one has discontinued this therapy and thus continues to remain exposed to the adverse effects of these drugs (Anastasilakis et al., 2018). Stimulating the function of osteoblast survival and function on the other hand is known as the osteoanabolic approach that addresses the issue of restoring the lost bone caused due to osteoporosis. In this regard, two peptide drugs (teriparatide and abaloparatide) targeting the type 1 parathyroid hormone receptor are in clinical use (Silva and Bilezikian, 2011). Because these are peptide drugs, the route of administration is thus parenteral (daily subcutaneous injection). These osteoanabolic drugs do not affect bone resorption, rather with time increase resorption that limits their use beyond 2 years. The only drug, romosozumab (a humanized antibody against sclerostin, a Wnt inhibitor), appears to have both osteoanabolic and anti-resorptive effects; however, these unique features are limited to 2 years, beyond which resorption catches up with the formation, leaving no net bone gain (Tabacco and Bilezikian, 2019). To summarize, anti-resorptive drugs inhibit both resorption and formation of bone and osteoanabolic drugs increase both formation and resorption of bone. This therapy increases bone formation and inhibit bone resorption and maintain this for long-term (well past 2 years) and is the most desirable anti-osteoporosis therapy, which is an important medical need.

Synthetic GCs are used as potent anti-inflammatory and immunosuppressive drugs, and the first line of therapies for inflammatory and autoimmune diseases. GC use beyond 6 months increases fracture risk even at normal BMD due to apoptosis of osteoblasts and osteocytes and an increase in osteoclast function (Ilias et al., 2000). When it comes to osteoporosis, it is said that there is no safe dose of GC. Moreover, unlike postmenopausal osteoporosis (PMO) where fracture risk is elevated in the cancellous parts (lumbar vertebra and femur neck), GC use increases the risk of fractures at the cortical bones as well. As in the case of PMO, BP is the first line of therapy for GIO (Compston, 2018). However, it should be noted that since the major aetiology of GIO is the loss of osteoblast/osteocyte viability and function and that BP is known to suppress osteoblast function; the use of BP in GIO is not an ideal therapy instead osteoanabolic therapy is more suitable. Moreover, high-dose GC through apoptosis induction effect on osteocytes causes osteonecrosis, and BPs are also reported to do the same, which suggests that the deterioration of bone quality due to GC could be worsened by BP (Lems and Saag, 2015). Despite greater suitability of teriparatide/abaloparatide for GIO, the limitation of time-dependent stimulation of resorption by these drugs will remain in this case (Taylor and Saag, 2019). Hence, an orally active therapy that stimulates bone formation with or without affecting resorption will be the desirable therapy for GIO.

GCs are widely used in rheumatoid arthritis (RA), which is a debilitating form of autoimmune disease. In RA, joints (cartilage and subchondral bones) are destroyed through a complex inflammatory process. Local (joint) and systemic inflammation in RA also induce bone loss due to the activation of osteoclasts by the inflammatory cytokines from T cells, macrophages, neutrophils, and fibroblast-like synovial (FLS) cells (Guo et al., 2018). Although GCs potently suppress the immune "flare-up" in RA and thereby delays the radiological progression of joint destruction, given its direct osteopenic effect as discussed in the preceding paragraph, protecting bone cells (particularly osteoblastic cells) from the apoptotic insult of GC is desirable, and an unmet medical need (Paolino et al., 2017; Burmester and Pope, 2017).

Although sharing the same symptoms of RA, osteoarthritis (OA), the major form of arthritis is caused due to very different conditions. The pathophysiology of OA includes the destruction of articular cartilage, structural changes in the underlying subchondral bone, and chronic inflammation of the synovium (Ashkavand et al., 2013). Since joint destruction also causes local inflammation, high doses of long-acting intra-articular GC injections are widely used in clinical practice, which could lead to the suppression of osteoblast function in the joint bones (Hartmann et al., 2016). Furthermore, a recent report showed that physical exercise is more effective in reducing joint pain and physical ability in OA patients than intra-articular GC injection, thereby suggesting that GC use in OA may not be necessary (Bhatia et al., 2013). Presently, there is no pharmacotherapy for OA and the primary course of medical management involves pain mitigation through analgesic drugs besides intra-articular GC injection for mitigating local inflammation.

In the following sections, we review the effects of CO extract, the fraction derived from the extract with enrichment of osteogenic compounds, development of a formulation to enhance the oral bioavailability of osteogenic compounds, and constituent compounds of CO in relation to fracture healing, osteoporosis and arthritis.

4.3 THE EFFECTS OF CO EXTRACT FRACTION AND FORMULATION

Since the nineteenth century, extracts of leaf and stem were used for accelerated fracture healing as a traditional medicine in Puttur (Andhra Pradesh, India) (Singh, 2017; Yadav et al., 2010). Pal et al. showed that ethanolic extracts of stem and leaf were effective for bone regeneration at fracture site in a femur osteotomy model. The stem extract of CO (CSE) was effective at 250 mg/kg versus 750 mg/kg dose of the leaf extract of CO (CLE) (Pal et al., 2019). From the butanolic fraction of CSE, ten compounds were isolated out of which six were found to be osteogenic through in vitro assessments of ALP activity and expression of osteogenic genes that included bone morphogenetic protein-2

(BMP-2), runt-related transcription factor-2 (Runx2) and type 1 collagen (col1). The osteogenic compounds were apigenin, luteolin, emodin, isovitexin, 4′,7′-dihydroxyflavone and 3′,4′,7′-trihydroxyflavone (Pal et al., 2019). Among the osteogenic compounds, isovitexin and 3′,4′,7′-trihydroxyflavone were found to upregulate all three osteogenic genes thus suggesting these two compounds to be more effective than the other four in stimulating bone formation. The butanolic fraction of CSE (CBE) increased bone regeneration in the femur osteotomy model at 100 mg/kg dose, which suggested enhancement of the osteogenic activity of CBE over CSE which was attributed by the enrichment in CBE of six osteogenic compounds described before (Pal et al., 2019).

In a widely used model of iatrogenic osteoporosis obtained by the administration of a corticosteroid, methylprednisolone (MP), CSE (250 mg/kg) and CBE (100 mg/kg) mitigated MP-induced bone loss and strength by osteogenic as well as anti-catabolic actions. Even at a 2.5-fold lesser dose than CSE, the osteogenic effect of CBE was significantly greater than CSE. Both CSE and CBE mitigated MP-induced loss of body weight and electrolyte imbalances with later being more effective (Pal et al., 2019). Through an LC-MS/MS method, apigenin, isovitexin, luteolin, emodin and trihydroxyflavone were detected in adult rat plasma after an oral administration of CSE (500 mg/kg) (Pal et al., 2019). These compounds were stable in simulated gastric and intestinal fluids but were metabolized in rat liver microsomes. The development of bioanalytical methods for pharmacokinetics and *in vitro* stability studies of osteogenic compounds will be useful for the phase 1 clinical trial.

Self-nano emulsifying drug delivery system (SEDDS) is an efficient mode for improving the bioavailability of poorly absorbed compounds often present in phytoextracts. A lipid-based SEDDS of CBE was found to enhance the bioavailabilities of apigenin, isovitexin, THF, luteolin and emodin along with the increase in the skeletal effect. However, CBE at 100 mg/kg dose increased osteogenic effect, the SEDDS formulated CBE achieved the same effect at 50 mg/kg (Pal et al., 2020). The study also found that MP treatment significantly suppressed osteocyte markers including dentin matrix acidic phosphoprotein 1 (DMP-1) and matrix extracellular phosphoglycoprotein (MEPE), and SEDDS formulated CBE maintained their expression (Pal et al., 2020). Muscle atrophy is another signature of GC treatment and we found that SEDDS-formulated CBE significantly improved muscle structure and prevented muscle atrophy. Reports also showed that SEDDS-formulated CBE did not alter the anti-inflammatory effect of MP (Pal et al., 2020). These studies established CO extract and its related formulation as an effective pharmacotherapy for the treatment of GC-induced osteo-sarcopenia.

4.4 EFFECTS OF COMPOUNDS DERIVED FROM CO EXTRACT

The phytochemicals such as apigenin, emodin, luteolin, kaempferol, quercetin and caryophyllene shown in Figure 4.1 and isolated from CO extract were evaluated for their osteogenic activity.

4.4.1 Apigenin

In human fetal bone marrow-derived MSCs, apigenin stimulated osteogenic differentiation and upregulated osteogenic genes (Runx2, osterix and osteopontin) through JNK and p38 MAPK pathways (Zhang et al., 2015). In osteoblasts, apigenin blocks the action of tumour necrosis factor α (TNFα) and interferon γ (IFNγ) by inhibiting the production of osteoclastogenic cytokines including interleukin-6 (IL-6), monocyte chemoattractant protein-1 (MCP-1), MCP-3, and regulated upon activation, normal T cell expressed (RANTES). In preadipocytes, apigenin inhibits adipocyte differentiation, production of osteoclastogenic cytokines MCP-1, MCP-3, enhanced the production of the osteogenic cytokine BMP-6. Apigenin inhibited the differentiation of pre-osteoclasts to mature osteoclasts and attenuated pit resorption by mature osteoclasts (Bandyopadhyay et al., 2006). These reports suggested that

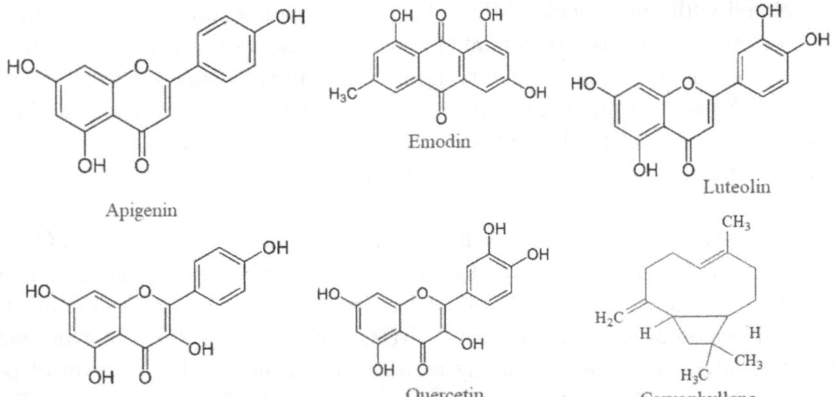

FIGURE 4.1 Structure of osteogenic active compounds derived from *Cassia occidentalis* extract.

apigenin has multiple effects on bone cells that could result in osteoanabolic as well as osteoprotective outcomes in diseases of bone loss.

The effect of apigenin in the condition of OVX-induced bone loss was assessed. Apigenin (10 mg/kg, p.o., 3X/week) given to osteopenic rats (in therapeutic mode) for 23 weeks restored trabecular bones at femur and tibia. The effect of apigenin at the trabecular bone was comparable to 17β estradiol (E2). Although it had no uterotrophic effect, the OVX-induced gain in body weight was significantly reduced by apigenin to the levels of the E2 group. Similar to E2, apigenin suppressed the bone turnover markers suggesting remodelling suppression as the mode of its action mechanism. Vertebral bone, consisting of trabecular bones and is a major site for fracture in post-menopausal osteoporosis has not been studied (Park et al., 2008). In OVX mice, apigenin (10 mg/kg) treatment for 4 weeks by intraperitoneal route (in preventive mode) protected from loss of femur trabecular bones. In this study, apigenin inhibited the differentiation of both murine osteoblasts and osteoclasts, which was reminiscent of the action of BPs, the potent anti-resorptive drug class. Unlike the inhibition of the OVX-induced increase in body weight of rats by apigenin, OVX mice given apigenin did not affect body weight gain (Goto et al., 2015). Thus, it appears that apigenin may have a species-specific effect on body weight, although the skeletal effect seems to be the same. Neither study assessed the functional impact of bone restoration/conservation through the assessment of bone strength. From these preclinical models of postmenopausal osteoporosis, it appears that apigenin could suppress the loss of BMD in postmenopausal women. Although there is no clinical study assessing such effect of apigenin, urinary apigenin levels were negatively correlated with femur neck BMD in healthy Korean postmenopausal women (Kim et al., 2002). Also, on a related note, assessment of urinary excretion of isoflavones as a marker of dietary intake in healthy postmenopausal Dutch women showed no preventive effect of phytoestrogens on cortical bone loss (Kardinaal et al., 1998). Therefore, there is a need for conducting high-quality clinical trials to determine the role of apigenin in postmenopausal osteoporosis.

However, the suppression of bone turnover markers by apigenin in OVX rats suggests the anti-resorptive mode of its action, another report showed that apigenin promoted repair of skull defect in the adult mouse. Apigenin increased osteoblastic differentiation of MSCs derived from bone marrow via the activation of the Wnt pathway assessed from nuclear localization of β-catenin (Jiang et al., 2019). Conversely, in human U2 OS, osteosarcoma cells, apigenin-induced apoptosis and at 2 mg/kg dose given parenterally for 30 days attenuated xenograft tumour growth in nude mice (Lin et al., 2012). From these two reports, it appears that the effects of apigenin vary between normal and malignant cells.

The 8,8″-biapigeninyl, a condensation product of two apigenin molecules that is abundantly present in *Cupressus sempervirens*, was found to be 10^4-fold more potent than apigenin in stimulating osteoblast differentiation. In OVX mice, 8,8″-biapigeninyl at 1 mg/kg oral dose significantly protected against bone loss by osteogenic and anti-resorptive mechanisms (Siddiqui et al., 2010). It should also be noted that apigenin-6C-glucopyranose (isovitexin) is ~3-fold more abundant than apigenin in the CO extract and is more potent than apigenin in cultured osteoblasts. Also, when CO extract was administered orally in rats, isovitexin showed better metabolic stability than apigenin. Although there is no report on in vivo skeletal effect of isovitexin, based on *in vitro* and metabolic stability data, it is surmised that this compound will have a better bone effect than apigenin.

Apigenin also has an immunomodulatory effect through which it is reported to mitigate RA. The antigen-presenting dendritic cells (DCs) contribute to the pathogenesis of RA by (a) infiltration of immature DCs into the synovial tissues and gain maturity in the inflammatory milieu by the action of proinflammatory cytokines including TNFα, IL-1β and IL-6 (Lutzky et al., 2007); (b) mature DC then migrate to the secondary lymphoid organs and present antigens to naïve T cells to cause T-cell activation and T-cell-mediated cytotoxicity (Wenink et al., 2008); and (c) mature DCs induce B-cell-mediated antibody production (Lebre et al., 2008). In DCs derived from mouse bone marrow (the CD11c+ cells), lipopolysaccharide (LPS), the TLR ligand increased surface expression of co-stimulatory molecules including CD40, CD80, CD86 and class II MHC, and apigenin blocked this effect as well as the phagocytic ability of DCs. Interestingly, apigenin suppressed the LPS-induced increase in TNFα but enhanced IL-1β, while had no effect on IL-6 and IL-17A. In a murine model of RA, apigenin (20 mg/kg i.p.) administered in a preventive mode mitigated the severity of the disease. Apigenin-treated mice had reduced maturation of DC and CXCR4 expression, the chemokine receptor involved in DC maturation. The migratory ability of DC in RA mice was also reduced by apigenin (Li et al., 2016).

4.4.2 Emodin

Emodin (1,3,8-trihydroxy-6-methyl anthraquinone) is an anthraquinone and has an anti-osteoclastic effect in mouse bone marrow-derived macrophages (BMM). Inhibition of osteoclastic differentiation effect of emodin was observed at 1 μM with maximum suppression at 10 μM. Significant suppression of pit resorption (in vitro osteoclast functional assay) was observed at 5- and 10 μM, which suggested that besides inhibiting osteoclast differentiation, emodin also suppressed osteoclast function. The anti-osteoclastic effect of emodin is mediated by the suppression of receptor activator of nuclear receptor κB ligand (RANKL)-induced nuclear factor κB (NFκB) signaling. In an acute inflammatory bone loss model induced by LPS in mouse, oral administration of emodin (50 mg/kg) suppressed bone loss and osteoclast number, which suggested that emodin inhibited bone resorption in vivo. Emodin also promoted osteoblast differentiation in vitro and increased the expression of Runx2 and osteocalcin in calvariae (Kim et al., 2014). Emodin stimulates osteoblast differentiation through the activation of the cellular prosurvival PI3K (phosphatidylinositol 3-kinase)-Akt (Protein Kinase B) pathway resulting in the downstream activations of JNK and p38 MAPK, and an increase in BMP-2 production by osteoblasts (Lee et al., 2008). From these reports, it appears that emodin is anti-resorptive and osteogenic.

Induction of osteoblast differentiation by emodin is associated with the upregulation of BMP-9, osterix, activin receptor-like kinase 1 (ALK1), smad 1, smad 9 and Msh homeobox 2 (Msx2) mRNA levels in osteoblasts. Emodin-induced osteogenic differentiation could be blocked by noggin, thus suggesting the role of BMP-9 as the mediator of this process as BMP-2 expression was unchanged by emodin. In OVX rats, emodin (100 mg/kg, route of administration unspecified) given for 12 weeks although inhibited tartrate-resistant acid phosphatase 5b (TRACP5b, the surrogate of osteoclast number) had no effect on preventing loss of bone volume and strength. A "low dose" E2 (50 μg/kg) also had no effect however, when combined with emodin, complete protection against OVX-induced trabecular osteopenia and loss of strength was observed. As 50 μg/kg E2 had no uterotrophic effect, a combination of low-dose E2 and emodin has been suggested for the treatment of postmenopausal osteoporosis (Chen et al., 2017).

Unlike in rat, where emodin had no effect on preventing the OVX-induced bone loss, in OVX mouse, emodin (100 mg/kg b.i.d. every 3 days) given for 3 months increased osteoblast number with the increase in Runx2 positive cells in the lumbar vertebra and contributed to increases in bone mass and improved microarchitecture. In mouse bone marrow, emodin inhibited the OVX-induced adipocyte number and fat tissue fraction. In cultures of MSCs derived from bone marrow, emodin induced proliferation as well as differentiation. Osteogenic genes including Runx2, osterix, osteocalcin, col1 and BMP-4 were upregulated by emodin. Furthermore, emodin suppressed the differentiation of MSCs to adipocytes with attendant decreases in adipogenic genes including peroxisome proliferator-activated receptor-gamma PPARγ, CCAAT/enhancer-binding protein alpha C/EBPα and adipocyte protein 2 (aP_2). Since increased adipogenesis significantly contributes to osteoclastogenic differentiation, emodin's suppression of adipogenesis is likely to inhibit osteoclastogenesis and ultimately resorption. However, this study did not assess resorption parameters in OVX mice (Yang et al., 2014).

Osteoporosis secondary to inflammatory bowel disease (IBD) is caused due to systemic inflammation. In adult male rats, IBD was created by 2.5% dextran sulfate (DS) in drinking water following which emodin treatment (30 mg/kg administered orally 3X/week for 9 weeks) was given. Compared to control rats (the disease group was given vehicle), emodin-treated rats had improved bone parameters (bone volume and strength) and reduced bone resorption markers (osteoclast number and serum C-terminal telopeptide/ CTX-1). Serum TNFα that was significantly elevated in the IBD group was suppressed by emodin (Luo et al., 2020). A standardized extract of *Polygonum multiflorum* (PM), containing emodin protected against prednisolone-induced loss of body weight and bone mass at the femur and lumbar vertebra. Microarchitecture was significantly preserved and bone strength showed an increasing trend in the PM group compared to the control group (Zhou et al., 2017). From these reports, it appears that emodin has a protective role in secondary osteoporosis.

NFκB is the major inflammatory molecule that not only supports osteoclastogenesis and bone resorption but also causes joint destruction in RA and OA. In a mouse model of collagen-induced RA, the so-called CIA model, emodin (10 mg/kg, i.p.) in a therapeutic mode (administered 25 days after collagen-II challenge) decreased the arthritis disease score, suppressed synovial inflammation and joint destruction. In the joints of CIA mice, the enhancement of NFκB signaling revealed by decreased cytoplasmic levels of inhibitor κB (IκB) and increased nuclear levels of p65 and p50 in the nuclear fraction was reversed by emodin treatment. Consequently, the CIA-induced elevations in inflammatory cytokines including TNFα, IL1β, RANKL and IL17 in the joints were decreased by emodin, as well as the serum levels of TNFα and IL1β. Furthermore, the matrix-degrading matrix metalloproteinases-1,3 (MMP1 and MMP3) levels that were increased in the joints of CIA mice were significantly suppressed by emodin treatment (Hwang, 2013).

In cultures of rat chondrocytes, emodin suppressed the cytotoxic impact of IL-1β, downregulated the expressions of MMP-1 and MMP-13 and inhibited the ERK and Wnt/β-catenin signaling (Liu et al., 2018). In murine chondrogenic cell line, ATDC5, emodin treatment caused the loss of cell viability at 15- and 20 μM. However, at 10 μM, emodin blocked the LPS-induced apoptosis of and production of proinflammatory cytokines (TNFα, IL-6 and MCP1) from ATDC5 cells. The pro-survival, anti-apoptotic and anti-inflammatory roles of emodin in ATDC5 were likely regulated by the long noncoding RNA, taurine upregulated gene 1-mediated notch and NFκB pathways (Liang and Ren, 2018). Given this action mechanism of emodin chondrocytes, a salutary role of this compound in OA is surmised.

4.4.3 Luteolin

In C3H10T1/2 cells, luteolin promotes osteogenic differentiation and in 3T3-L1 cells, it inhibits adipogenic differentiation. These effects are mediated by heat shock protein, Dnajb1 (DnaJ Hsp40) (Kwon et al., 2016). In human osteogenic sarcoma cell line, Saos2, luteolin, and its 8-*C*-glucopyranose analog orientin increased mineral content upon the induction of differentiation in the presence of β-glycerophosphate. At >10 μM, luteolin decreased the mineral content likely due to a pro-oxidant impact. At 50 μM, luteolin had a cytotoxic effect on osteoblasts as assessed by LDH release. Both luteolin and orientin inhibited the production of pro-inflammatory cytokines (TNFα and IL-6) from osteoblasts as well as sclerostin. Suppression of sclerostin by luteolin is likely to promote the osteogenic Wnt signaling (Nash et al., 2015). Furthermore, luteolin protected MC3T3-E1 cells against oxidative damage caused by H_2O_2 and menadione (Fatokun et al., 2015). From these reports, it appears that luteolin promotes osteoblast function and inhibits osteoclast function.

In OVX mice, luteolin at 20 mg/kg dose completely protected against loss of bone mass, bone strength and appearance of poor microarchitecture at the femur and lumbar spine. This effect was mediated by luteolin's inhibitory function on osteoclast formation (Kim et al., 2011). All bone parameters at both cancellous and cortical sites were protected against dexamethasone-induced osteopenia by 100 mg/kg luteolin in adult rats. High-dose dexamethasone causes loss of osteoblast viability and function, and luteolin prevented osteoblasts from such effects of the corticosteroid. Luteolin also inhibited ROS production from osteoblasts in response to dexamethasone (a GC) that inhibits osteoblast functions. Furthermore, an increased osteoprotegerin (OPG)/RANKL ratio in osteoblasts in response to luteolin suggested suppression of osteoclast formation and function. The osteogenic effect of luteolin was mediated by ERK and Wnt/β-catenin pathways (Jing et al., 2019). In human periodontal ligament cells which show some features of osteoblasts, luteolin at 1 μmol/L stimulated osteogenic differentiation, by Wnt/β-catenin mediated mechanism (Quan et al., 2019). In experimentally induced periodontitis in rats, luteolin at 50- and 100 mg/kg oral doses prevented bone loss by increasing osteoblast cell counts, and BMP-2 expressions and decreasing inflammation, osteoclast cell counts, RANKL/OPG ratio, MMP-8 activity and iNOS expressions (Balci Yuce et al., 2019).

Luteolin also modulates chondrocyte function. In chondrocytes derived from knee joints of rats, luteolin suppressed IL-1β-induced inflammatory mediators including nitric oxide (NO), prostaglandin E2 (PGE2), TNFα, and matrix proteases including MMP-1, MMP-2, MMP-3, MMP-8, MMP-9 and MMP-13. These effects are likely mediated by the downregulation of

IL-1β-induced NFκB. Luteolin also reversed the IL-1β-mediated downregulation of collagen II protein. Accordingly, in the monosodium iodoacetate (MIA)-induced model of OA, luteolin at 10 mg/kg oral dose reduced the severity of articular cartilage damage and significantly increased collagen II staining compared to control OA rats (Fei et al., 2019).

Luteolin has an immunomodulatory function that could beneficially impact RA. For example, luteolin (1 mg/kg) in combination with N-palmitoylethanolamide (PEA), an endogenous fatty acid amide belonging to the family of the N-acylethanolamines, significantly ameliorated the clinical signs of RA (erythema and oedema in hind paw) and pain and locomotor activity. Although PEA alone was effective in improving these parameters, the impact of the combination was better. Luteolin alone (1 mg/kg) had no effect. The combination therapy also inhibited neutrophil infiltration in the joint of RA mice as well as decreased the levels of inflammatory chemokines, macrophage inflammatory protein (MIP-1α) and MIP-2. Consequently, free radicals and oxidant molecules nitrotyrosine, a marker of nitrosative injury was increased in the joints of RA mice but significantly reduced in the RA mice treated with the combination of PEA and luteolin. The circulating levels of TNFα, IL-1β and IL-6 were also suppressed in the RA group receiving the combination treatment compared to the disease (RA) group. Moreover, thiobarbituric acid-reactant substance, an indicator of lipid peroxidation, was increased in the plasma of RA group, which was suppressed in the combination group. Thus, luteolin in combination with PEA improved experimental RA in mice by anti-inflammatory and analgesic mechanisms (Impellizzeri et al., 2013).

Luteolin has also been reported to regulate neutrophil function. Neutrophils importantly contribute to RA pathogenesis by promoting the loss of immune tolerance and increasing synovial joint inflammation. Neutrophils also cause NET formation in the synovium that then externalizes citrullinated proteins which could potentially act as autoantigens, thereby triggering autoimmune joint destruction cascades. Neutrophils also interact with synovial fibroblast (SF)-like cells to induce inflammatory response through MHC-dependent antigen presentation and formation of autoantibodies. In activated human neutrophils, luteolin was reported to inhibit (a) superoxide anion and ROS generation, (b) degranulation thereby suppressing elastase release, (c) NET formation thereby suppressing the release of decondensed chromatin coated with inflammatory cytokines and (d) chemotaxis likely via the suppression of the surface adhesion protein CD11b. These actions are mediated by MEK-1- and Raf1-dependent mechanism. Furthermore, in an acute model of paw swelling in mice induced by CFA, luteolin (50 mg/kg, single injection) attenuated paw inflammation, ROS production and polymorphonuclear granulocytes in the paw (Yang et al., 2018). In a chronic model of paw swelling, luteolin at a minimum dose of 10 mg/kg suppressed paw oedema and inflammation with concomitant decreases in inflammatory cytokines (IL-1β, IL-6, IL-17 and TNFα). In the inflamed paws, luteolin inhibited the inflammatory P2X4 (a membrane cation-selective channel and NLRP1 inflammasome) (Shi et al., 2015).

FLS isolated from rats with experimental RA displayed suppressed proliferation, production of inflammatory cytokines, and the production of MMP-1 and MMP-3 in response to luteolin. Besides, luteolin mitigated the TNFα-induced proliferation and production of inflammatory cytokines and MMPs from FLS. Attenuation of the proliferation of FLS by luteolin is caused by a G2/M delay, i.e. entry into mitosis by the upregulation of the cell cycle inhibitor p21, and blocking the activation of the ERK and PI3K-Akt pathways (Hou et al., 2009). These reports describe the protective roles of luteolin in RA through various anti-inflammatory and non-anti-inflammatory mechanisms.

4.4.4 Caryophyllene

Caryophyllenes are natural sesquiterpenes present in the essential oils of many medicinal plants including CO extract. In MSCs derived from mouse bone marrow, β-caryophyllene between 0.1 and 100 μM range promotes osteoblastic differentiation but inhibits adipocytic and osteoclastic differentiation (Yamaguchi and Levy, 2016). In MC3T3-E1 cells, trans-caryophyllene promotes differentiation assessed by ALP activity, collagen production and osteocalcin release, and these effects were mediated by CB2R, a cannabinoid receptor isoform. Trans-caryophyllene also protected cells against antimycin-induced loss viability of MC3T3-E1 cells likely due to the reduction of ROS production due to the activation of anti-oxidant machinery including increased GSH levels and catalase activity. Furthermore, the endogenous antioxidant stress protectant proteins including Nrf2 and heme oxygenase 1 (HO-1) that were downregulated by antimycin A were maintained in MC3T3-E1 cells with trans-caryophyllene co-treatment (Shan et al., 2017).

In experimental RA mice induced by collagen antibody, β-caryophyllene at 10 mg/kg oral dose significantly reduced the severity of arthritis score and protected joints from destruction. Both serum and joint inflammatory cytokines (TNFα, IL-1β, IL-6 and IL-13) were reduced compared to RA control mice. In the joints of RA mice, β-caryophyllene decreased (a) matrix proteases (MMP-3 and MMP-9) and (b) pro-inflammatory biomolecules including NFκB and cyclooxygenease 2 (COX2). Activation of NFκB suppresses the expression of peroxisome proliferator-activated receptor-γ coactivator 1α (PGC-1α) in inflammatory cells and low levels of PGC1-α results in NFκB activation, and together these events represent a vicious cycle triggering inflammatory insult to the joints (Rius-Pérez et al., 2020). The expression of PPARγ in monocytes of RA patients has an inverse association with RA disease activity, which suggested that overexpression of PPARγ may have anti-RA effects (Ma et al., 2019). β-caryophyllene prevented the decreased expression of PPARγ and PGC1α levels in the joints of RA mice. An inhibitor of CB2R, AM630 blocked the effects of β-caryophyllene on NFκB, PPARγ and PGC-1α. Because PPARγ activation is downstream of CB2R activation (O'Sullivan, 2016), it is proposed that the anti-RA effect of β-caryophyllene is mediated by the CB2R pathway (Irrera et al., 2019). Treatment of β-caryophyllene in prophylactic mode at 215- and 430 mg/kg doses by oral gavage to rats with paw inflammation model (adjuvant-induced inflammation model induced by Freund's adjuvant) mitigated the articular and systemic inflammation in addition to attenuation of plasma and liver oxidative stress by upregulating the liver anti-oxidant machinery (Ames-Sibin et al., 2018).

4.4.5 Kaempferol

The literature is rather replete with the effects of kaempferol in bone and joint, and has recently been reviewed exhaustively (Wong et al., 2019). Here, we will give a summary of the major findings. Kaempferol favours osteoblast differentiation with increases in osteogenic genes such as BMP-2, osterix and type I collagen. Through proteomics-based profiling, cytokeratin-14 and HSP-70 were upregulated, whereas caldesmon and aldose reductase were downregulated in rat calvarial osteoblasts by kaempferol (Kumar et al., 2010). The differentiation of osteoblast by kaempferol was mediate by cytokeratin-14 through the participation of AMPK and mTOR signaling (Khedgikar et al., 2016). Because silencing HSP-70 mRNA in human MSCs suppressed osteogenic and chondrogenic differentiation (Li et al., 2018), it is surmised that the osteogenic differentiation of kaempferol was mediated by HSP-70. However, the implication of the downregulation of caldesmon and aldose reductase by kaempferol in regulating osteoblast function is not known.

The anti-oxidant effect of kaempferol in osteoblasts has been studied. 2-Deoxy-D-ribose (dRib) is a sugar with a high reducing capacity and induces oxidative stress and kaempferol mitigates dRib-induced oxidative stress in MC3T3-E1 cells by reducing malondialdehyde content (Suh et al., 2009). Kaempferol is also known to activate both ERα- and ERβ transactivation and promote the expression of osteogenic genes (Guo et al., 2012; Tang et al., 2008). In MC3T3-E1 cells, induction of differentiation by kaempferol was accompanied by autophagy assessed by the expression of beclin-1, sequestosome-1, and conversion of LC3-II to LC3-I (Kim et al., 2016).

Kaempferol also protects osteoblasts against dexamethasone-induced apoptosis through the activation of the ERK pathway (Adhikary et al., 2018). Together the data suggest that kaempferol promotes osteogenic differentiation, protects against oxidative stress, and drug-induced apoptosis. The signaling mechanisms are varied including ER, BMP, mTOR and autophagy pathways.

Besides osteoblasts, chondrogenic differentiation is also supported by this flavonoid which is associated with increased synthesis of matrix proteins including types I and X collagen and proteoglycans. Conversely, osteoclast differentiation and function are suppressed by kaempferol through reduced expression of nuclear factor of activated T-cells, cytoplasmic 1 (NFATc1) and c-Fos. Kaempferol also blocked the TNF-induced production of osteoclastogenic cytokines such as IL-6 and MCP-1 from osteoblasts (Pang et al., 2006). In mature osteoclasts, kaempferol induced apoptosis and in vitro resorption through the participation of alleviation of intracellular ROS and ER signaling (Wattel et al., 2003). Inhibition of osteoclastogenesis by kaempferol was associated with the downregulation of autophagy-related protein, p62/SQSTM1 (Kim et al., 2018). These in vitro observations suggested kaempferol as an osteoanabolic and osteoprotective flavonoid.

Although kaempferol has a poor oral bioavailability yet in OVX osteopenic rats, the flavonoid at 5 mg/kg daily oral dose prevented trabecular bone loss, improved vertebral strength and suppressed bone turnover. When the bioavailability pitfall of kaempferol was bypassed by the application of a novel layer-by-layer polyelectrolyte nano-matrix resulted in a ~70% increase in the bioavailability of kaempferol over the unformulated kaempferol and correspondingly enhanced the significantly enhanced skeletal impact in OVX rats (Kumar et al., 2012).

4.4.6 Quercetin

Supplementation of dietary quercetin (2.5% in diet) to OVX mice for 4 weeks that achieved its plasma levels to 7 μM/L significantly improved BMD and bone parameters over the OVX mice with a vehicle. Although quercetin inhibited osteoclast function in vitro, it had no effect on increased osteoclast number and surface area caused due to OVX. Quercetin had no uterotrophic effect and showed estrogen receptor (ERα and ERβ) agonistic effect by luciferase-based receptor transactivation assay. From these findings, it appears that quercetin protects bones from estrogen deficiency-induced loss by an estrogen-independent mechanism likely involving osteogenic mechanism (Tsuji et al., 2009). Quercetin (150 mg/kg, p.o.) also protected female rats against GC-induced bone loss. Quercetin augmented femoral trabecular and cortical thickness, cortical bone strength, osteoblast number and serum osteocalcin compared to GC group. In this study, quercetin was found to be primarily bone anabolic as bone resorption by GC was not suppressed as efficiently as alendronate. Furthermore, the bone anabolic response (measured by increased serum osteocalcin) of quercetin in GIO rats remains intact in presence of alendronate, which raised the possibility of using this compound in conjunction with alendronate (Derakhshanian et al., 2013).

Although orally administered quercetin was found osteogenic in OVX and GC-treated rats, yet the flavonoid like all others in this category suffers from poor oral bioavailability. To circumvent this limitation, the delivery of quercetin through the skin was tried. In OVX rats, phytosome-based encapsulation of quercetin decreased bone remodelling rate (assessed by serum ALP) over the free quercetin which suggested the potential application of such a strategy for enhancing the therapeutic efficacy of quercetin (Abd El-Fattah et al., 2017). Moreover, skin permissive quercetin-loaded transferosomes in chitosan films were designed, which when applied on hind limb of rats treated with GC, displayed reduced osteoblast apoptosis, a decline in osteoclastogenesis, and increases in femoral bone length, weight, thickness and density over the GC alone group (Pandit et al., 2020). These reports raise the possibility of topical use of quercetin for the treatment of diseases related to bone loss.

Besides the preventive/therapeutic effects of quercetin in osteoporotic conditions (OVX-induced or GC-induced), a large number of reports exist on the joint-protective effects of quercetin in a variety of preclinical models of RA and OA. In human RA-derived FLS cells which are the major pathogenic mediator of RA, quercetin suppressed IL-17-induced RANKL production and phosphorylation of mTOR, ERK and IκB-α, which suggested mitigation of inflammatory and bone resorptive outcomes. Furthermore, the IL-17-induced osteoclastogenesis from human CD14+ monocytes was attenuated by quercetin. When the monocytes were cocultured with RA-FLS prestimulated with IL-17, the presence of quercetin significantly attenuated the osteoclast differentiation. Also, the osteoclastic differentiation of CD14+ cells in the presence of Th17 cells was suppressed. Finally, quercetin decreased Th17 cells differentiation without affecting the regulatory T (Treg) cells (Kim et al., 2019). In FLS of RA patients, quercetin induced apoptosis, through the upregulation of a long non-coding RNA (lncRNA), metastasis-associated lung adenocarcinoma transcript 1 (MALAT1) (Pan et al., 2016). These reports suggest that quercetin targets FLS, the major cell type involved in RA pathogenesis from multiple directions and thus showing potential in the treatment of this disease. Accordingly, in a double-blind randomized clinical trial in women with RA, 500 mg quercetin supplementation daily for 8 weeks significantly decreased early morning stiffness, morning pain, after-activity pain, disease activity score-28 (DAS-28) and plasma hs-TNFa compared to the placebo group (Javadi et al., 2017).

Several preclinical studies demonstrated the ameliorating effect of quercetin in RA through multiple mechanisms. In collagen II-induced RA in rats, 150 mg/kg quercetin orally ameliorated the arthritic index score, decreased the serum inflammatory cytokines (TNFα, PGE2, IL-1β, IL-6, IL-17 and IL-21) and increased the anti-inflammatory cytokine, IL-10 compared to controls. Quercetin restored Th-17/Treg balance, inhibits NLRP3 inflammasome activation in the joint and increased the expression of HO-1, the protein that inhibits synovial inflammation (Yang et al., 2018). At a lower dose, i.e. 30 mg/kg quercetin through oral route ameliorated CII-induced RA in mice assessed from the arthritic score, joint histology, serum levels of inflammatory cytokines and MMPs, and the compound ameliorated the weight loss caused by methotrexate without affecting the strong anti-RA impact of methotrexate. It is thus conceivable that quercetin in conjunction with methotrexate could protect against the adverse effects of this disease-modifying anti-arthritic drug (DMARD) (Haleagrahara et al., 2018). In rats with adjuvant-induced arthritis, 150 mg/kg quercetin was found to reduce the plasma inflammatory markers (IL-1β, CRP and MCP-1) and anti-oxidant power, improved arthritic disease index, decreased the joint levels of NFκB and increased HO-1 (Gardi et al., 2015).

Purinergic signalling modulates inflammatory and immune responses via extracellular adenine, adenosine triphosphate (ATP), adenosine diphosphate (ADP), adenosine monophosphate (AMP) and adenosine. During inflammation, these extracellular biomolecules of purinergic signaling are regulated in immune cells by membrane-bound enzymes including ectonucleoside triphosphate diphosphohydrolase (E-NTPDase) that breaks down ATP and ADP to AMP (Yegutkin, 2008) and ectoadenosine deaminase (E-ADA) that is responsible for the removal of extracellular by deamination of adenosine and 2′-deoxyinosine (Latini and Pedata, 2001). The activity of E-NTPDase was increased and E-ADA was decreased in the lymphocytes of complete Freund's adjuvant (CFA)-induced arthritis, and oral quercetin treatment to arthritic rats reversed these changes along with the mitigation of arthritic scores. Serum adenosine level was higher in the arthritis rats compared to disease-free rats and quercetin significantly decreased this level. This report showed that the immunomodulatory effect of quercetin in arthritis involves the modulation of purinergic signaling (Saccol et al., 2019).

Besides systemic administration, topical application of quercetin has also been developed for RA treatment. A nanoemulsion of quercetin improved its skin permeability and anti-arthritic activity in a CFA-induced RA model in rats. This is a promising approach for topical application of quercetin to mitigate RA-related disease activity and reduce the need for NSAIDs, whose long-term use has a major side effect of GI ulcers (Gokhale et al., 2019).

Concerning OA, the chondroprotective effect of quercetin has been assessed. Quercetin protected rat articular chondrocytes from IL-1β-induced loss of cell viability and degradation of matrix proteins including col2 and aggrecan, and glycosaminoglycans. Quercetin also suppressed the levels of IL-1β-induced inflammatory and catabolic mediators in chondrocytes including inducible NO synthase (iNOS), COX2, MMP-13, and a disintegrin and metalloproteinase with thrombospondin motifs (ADAMTS)-4 by inhibiting the Akt and NFκB signaling. Although quercetin had no effect on the expression of M1 phenotype genes (mostly pro-inflammatory), it significantly upregulated the M2 phenotype genes including arginase 1, chitinase 3-like protein 3 and mannose receptor, which suggested that quercetin favours polarization of synovial macrophages into M2 phenotype which in turn provides a prochondrogenic environment required for remodelling of injured cartilage (Hu et al., 2019).

In rat articular chondrocytes, quercetin protected against Tert-butyl hydroperoxide-induced apoptosis by inhibiting oxidative stress and endoplasmic reticulum (ER) stress (downregulation of CHOP and GRP78). Quercetin also upregulated Sirt1, which maintains ER homeostasis and AMPK activation. Destabilization of the medial meniscus by transection of medial meniscotibial ligament caused OA-like condition in rats, and in these rats, quercetin (100 mg/kg, i.p.) protected against cartilage degradation accompanied by local suppression of ER stress and activation of Sirt1 and AMPK (Feng et al., 2019).

Using an OA model in rabbit achieved by medial parapatellar incision, quercetin (25 mg/kg, p.o.) was found to downregulate the expression of MMP-13 and upregulate the expression of tissue inhibitor of metalloproteinase-1 (TIMP-1) in serum, synovial fluid and synovium. Further, quercetin increased the production of superoxide dismutase (SOD) in serum and synovial fluid of OA rabbits. Taken together, through these mechanisms, quercetin improved OA-like degeneration in rabbits and the effect was comparable to celecoxib, and NSAID (Wei et al., 2019).

Conclusions 5

Analytical methods were developed for qualitative and quantitative analysis of compounds by HPLC-QTOF-MS and UPLC-QqQLIT-MS/MS in multiple reaction monitoring (MRM) acquisition mode respectively in different plant parts of *Cassia* species. Qualitative analysis resulted identification of twenty four compounds while eighteen compounds were also successfully quantified. The developed quantitative method was also validated according to ICH guidelines. Quantitative analysis revealed highest content of anthraquinones in seed/pulp of *C. fistula* and *C. auriculata*, in comparison to other parts and other *Cassia* species. Hence, this study clearly suggested the suitability of these parts as raw material in the manufacturing of *Cassia* based products. Six osteogenic marker compounds were standardized in extract and identified five of those in plasma after oral dosing of the extract to rats. A hydroalcoholic extract of leaf and stem of CO for up to 5 g/kg was found to be well-tolerated and safe in acute toxicity study and up to 2.5 mg/kg in 30-day subacute (repeat dose) study in rats (21571057). Compounds isolated from various parts of CO mostly fall within flavonoids, anthraquinones and essential oils. In CSE, ten of these were identified out of which six including apigenin, isovitexin, emodin, luteolin, 4′,7′-dihydroxyflavone and 3′,4′,7-trihydroxyflavone showed osteogenic activity in vitro. Besides promoting osteoblast function, these compounds and others present in CO (kaempferol, caryophyllenes and quercetin) have been reported to regulate the function of other bone cells including osteoclasts, chondrocytes and mesenchymal stem cells (MSCs). Consequently, compounds present in leaf and stem prevent osteoporosis, promote fracture healing and mitigate arthritis and joint degeneration. Thus, it is conceivable that CSE would protect against OVX-induced bone loss and RA. Action mechanisms include protection of osteoblasts and chondrocytes from oxidative damage, suppression of NFkapaB signalling and actions of pro-inflammatory cytokines, and regulation of a variety of MAPKs. Plant samples were also screened for their *in-vitro* anti-proliferative activity, in which results were widely varied Principal component analysis showed remarkable differences in the distribution of the chemical markers in various plant parts of *Cassia* species.

DOI: 10.1201/9781003186281-5

Conclusions

Analytical methods were developed for qualitative and quantitative analysis of compounds by HPLC-QTOF-MS and UPLC-QqQLIT-MS/MS in multiple reaction monitoring (MRM) acquisition mode respectively in different plant parts of *Cassia* species. Qualitative analysis resulted in identification of twenty-four compounds while eighteen compounds were also successfully quantified. The developed quantitative method was also validated according to ICH guidelines. Quantitative analysis revealed highest content of anthraquinones in seed/pulp of *C. fistula* and *C. auriculata*, in comparison to other parts and other *Cassia* species. Hence, this study clearly suggested the suitability of these parts as raw material in the manufacturing of *Cassia*-based products. Six [illegible] marker compounds were standardized in extract and identified five [illegible] in plasma after oral dosing of the extract to rats. A hydroalcoholic extract of leaf and stem of *C. occidentalis* up to [illegible] was found to be well tolerated and safe in acute toxicity study and up to [illegible] in 90-day subacute [illegible] dose) study in rats (2457/08 [illegible]). Compounds isolated from various parts of *C. occidentalis* [illegible] flavonoids, anthraquinones and [illegible] [illegible] CSE. Few of these were identified out of which [illegible] apigenin, [illegible] luteolin, 4',7-dihydroxyflavone and 3',4',7-trihydroxyflavone showed osteogenic activity in vitro. Besides, [illegible] [illegible] of the compounds and others [illegible] carbohydrates and [illegible] have been [illegible] to [illegible] other [illegible] [illegible] cells [illegible] [illegible] [illegible] osteoporosis [illegible] and [illegible] generation [illegible] that CSE [illegible] against [illegible] induced bone [illegible] [illegible] production of osteoblasts and [illegible] [illegible] [illegible] signalling and actions of pro-inflammatory cytokines, and modulation of a [illegible] of MAPKs. [illegible] samples were also [illegible] for their anti-proliferative activity [illegible] [illegible] widely [illegible]. Principal component analysis showed remarkable differences in the distribution of the chemical markers in various plant parts of *Cassia* species.

References

Abd El-Fattah, Abeer I., Mohamed M. Fathy, Zeinab Y. Ali, Abd El Rahman A. El-Garawany, Ehsan K. Mohamed. 2017. "Enhanced Therapeutic Benefit of Quercetin-Loaded Phytosome Nanoparticles in Ovariectomized Rats." *Chemico-Biological Interactions* 271 (June):30–38. 10.1016/j.cbi.2017.04.026.

Adaramoye O.A., Sarkar J., Singh N., Meena S., Changkija B., Yadav P.P., Kanojiya S., Sinha S. 2011. "Antiproliferative Action of *Xylopia aethiopica* Fruit Extract on Human Cervical Cancer Cells." *Phytotherapy Research* 25:1558–1563.

Adhikary, Sulekha, Dharmendra Choudhary, Naseer Ahmad, Anirudha Karvande, Avinash Kumar, Venkatesh Teja Banala, Prabhat Ranjan Mishra, Ritu Trivedi. 2018. "Dietary Flavonoid Kaempferol Inhibits Glucocorticoid-Induced Bone Loss by Promoting Osteoblast Survival." *Nutrition* 53 (September):64–76. 10.1016/j.nut.2017.12.003.

Ali M.S., Azhar I., Amtul Z., Ahmad V.U., Usmanghani K. 1999. "Antimicrobial Screening of Some Caesalpiniaceae." *Fitoterapia* 70:299–304.

Ames-Sibin, Ana P., Camila L. Barizão, Cristiane V. Castro-Ghizoni, Francielli M.S. Silva, Anacharis B. Sá-Nakanishi, Lívia Bracht, Ciomar A. Bersani-Amado, Maria R. Marçal-Natali, Adelar Bracht, Jurandir F. Comar. 2018. "β-Caryophyllene, the Major Constituent of Copaiba Oil, Reduces Systemic Inflammation and Oxidative Stress in Arthritic Rats." *Journal of Cellular Biochemistry* 119 (12):10262–10277. 10.1002/jcb.27369.

Anastasilakis, Athanasios D., Stergios A. Polyzos, Polyzois Makras. 2018. "Denosumab vs Bisphosphonates for the Treatment of Postmenopausal Osteoporosis." *European Journal of Endocrinology*. BioScientifica Ltd. 10.1530/EJE-18-0056.

Anonymous. "The Wealth of India: A Dictionary of Indian Raw Materials & Industrial Products, Publications & Information Directorate." *Council of Scientific and Industrial Research* 3:368–370.

Arya, Vedpriya, Sanjay Yadav, Sandeep Kumar, J.P. Yadav. 2010. "Antimicrobial Activity of *Cassia occidentalis* L (leaf) Against Various Human Pathogenic Microbes." *Life Sci Med Res* 9 (1):e12.

Ashkavand, Zahra, Hassan Malekinejad, Bannikuppe S. Vishwanath. 2013. "The Pathophysiology of Osteoarthritis." *Journal of Pharmacy Research* 7 (1):132–138. 10.1016/j.jopr.2013.01.008.

Awomukwu, Daniel Azubuike, Biolouis Nyananyo. 2015. "Comparative Chemical Constituent of Some Cassia Species and Their Pharmacognistic Importance in Southeastern Nigeria." *Science Journal of Chemistry* 3(3): 40–49.

Balci Yuce, Hatice, Hulya Toker, Ali Yildirim, Mehmet Bugrul Tekin, Fikret Gevrek, Nilufer Altunbas. 2019. "The Effect of Luteolin in Prevention of Periodontal Disease in Wistar Rats." *Journal of Periodontology* 90 (12):1481–1489. 10.1002/JPER.18-0584.

Bandyopadhyay, Sanghamitra, Jean Marc Lion, Romuald Mentaverri, Dennis A. Ricupero, Said Kamel, Jose R. Romero, Naibedya Chattopadhyay. 2006. "Attenuation of Osteoclastogenesis and Osteoclast Function by Apigenin." *Biochemical Pharmacology* 72 (2):184–197. 10.1016/j.bcp.2006.04.018.

Bhagat, Madhulika, Ajit Kumar Saxena. 2010. "Evaluation of *Cassia occidentalis* for In Vitro Cytotoxicity Against Human Cancer Cell Lines and Antibacterial Activity." *Indian journal of pharmacology* 42 (4):234.

Bhakta T., Banerjee S., Mandal S.C., Maity T.K., Saha B.P., Pal M. 2001. "Hepatoprotective Activity of *Cassia fistula* Leaf Extract." *Phytomedicine* 8:220–224.

Bhatia, Dinesh, Tatiana Bejarano, Mario Novo. 2013. "Current Interventions in the Management of Knee Osteoarthritis." *Journal of Pharmacy and Bioallied Sciences*. Wolters Kluwer – Medknow Publications. 10.4103/0975-7406.106561.

Bhope S.G., Kuber V.V., Nagore D.H. 2010. "Validated HPTLC Method for Simultaneous Quantification Of Sennoside A, Sennoside B, and Kaempferol in *Cassia fistula* Linn." *Acta Chromatographica* 22:481–489.

Burmester, Gerd R., Janet E. Pope. 2017. "Novel Treatment Strategies in Rheumatoid Arthritis." *The Lancet*. Lancet Publishing Group. 10.1016/S0140-6736(17)31491-5.

Cannarella, Rossella, Federica Barbagallo, Rosita A. Condorelli, Antonio Aversa, Sandro La Vignera, Aldo E. Calogero. 2019. "Osteoporosis from an Endocrine Perspective: The Role of Hormonal Changes in the Elderly." *Journal of Clinical Medicine* 8 (10):1564. 10.3390/jcm8101564.

Chandra, Preeti, Renu Pandey, Brijesh Kumar, Mukesh Srivastva, Praveen Pandey, Jayanta Sarkar, Bhim Pratap Singh. 2015. "Quantification of Multianalyte by UPLC–QqQLIT–MS/MS and In-Vitro Anti-Proliferative Screening in Cassia Species." *Industrial Crops and Products* 76:1133–1141.

Chaudhari S.S., Chaudhari S.R., Chavan M.J. 2012. "Analgesic, Anti-Inflammatory and Anti-Arthritic Activity of *Cassia uniflora* Mill." *Asian Pacific Journal of Tropical Biomedicine* 2:S181–S186.

Chauhan K.N., Patel M.B., Valera H.R., Patil S.D., Surana S.J. 2009. "Hepatoprotective Activity of Flowers of *Cassia auriculata* R. Br. Against Paracetamol Induced Liver Injury." *Journal of Natural Remedies* 9:85–90.

Chen X., Zhang S., Chen X., Hu Y., Wu J., Chen S., Chang J., Wang G., Gao Y. 2017. "Emodin Promotes the Osteogenesis of MC3T3-E1 Cells via BMP-9/Smad Pathway and Exerts a Preventive Effect in Ovariectomized Rats." *Acta Biochimica et Biophysica Sinica* 49 (10). 10.1093/ABBS/GMX087.

Chewchinda S., Sakulpanich A., Sithisarn P., Gritsanapan W. 2012. "HPLC Analysis of Laxative Rhein Content in *Cassia fistula* Fruits of Different Provenances in Thailand." *Thai Journal of Agricultural Science* 45:121–125.

Chewchinda S., Sithisarn P., Gritsanapan W. 2014. "Rhein Content in *Cassia fistula* Pod Pulp Extract Determined by HPLC and TLC-Densitometry in Comparison." *J Health Research* 28(6): 409–411.

Chewchinda S., Wuthi-udomlert M., Gritsanapan W. 2013. "HPLC Quantitative Analysis of Rhein and Antidermatophytic Activity of *Cassia fistula* Pod Pulp Extracts of Various Storage Conditions." *BioMed Research International* 2013: 1–8.

Clark, Dan, Mary Nakamura, Ted Miclau, Ralph Marcucio. 2017. "Effects of Aging on Fracture Healing." *Current Osteoporosis Reports*. Current Medicine Group LLC 1. 10.1007/s11914-017-0413-9.

Clarke, Bart L., Sundeep Khosla. 2010. "Physiology of Bone Loss." *Radiologic Clinics of North America*. NIH Public Access. 10.1016/j.rcl.2010.02.014.

Compston, Juliet. 2018. "Glucocorticoid-Induced Osteoporosis: An Update." *Endocrine*. Humana Press Inc. 10.1007/s12020-018-1588-2.

Cowen, Laura. 2019. "Global RA Burden 'Significant, yet Under-Recognized' | Rheumatology.Medicinematters.Com." *Medicine Matters*. https://rheumatology.medicinematters.com/rheumatoid-arthritis-/epidemiology-/global-ra-burden-significant-yet-under-recognized/17178360.

Danish M., Singh P., Mishra G., Srivastava S., Jha K.K., Khosa R.L. 2011. "*Cassia fistula* Linn. (Amulthus)—An Important Medicinal Plant: A Review of Its Traditional Uses, Phytochemistry and Pharmacological Properties." *Journal of Natural Product and Plant Resources* 1:101–118.

Das, Gouranga, Durbadal Ojha, Bolay Bhattacharya, Monisankar Samanta, Soma Ghosh, Suman Datta, Amalesh Samanta. 2010. "Evaluation of Antimicrobial Potentialities of Leaves Extract of the Plant *Cassia tora* Linn.(Leguminosae/Caesalpinioideae)." *Journal of Phytology* 2(5): 64–72.

Derakhshanian, Hoda, Mahmoud Djalali, Abolghassem Djazayery, Keramat Nourijelyani, Sajad Ghadbeigi, Hamideh Pishva, Ahmad Saedisomeolia, Arash Bahremand, Ahmad Reza Dehpour. 2013. "Quercetin Prevents Experimental Glucocorticoid-Induced Osteoporosis: A Comparative Study with Alendronate." *Canadian Journal of Physiology and Pharmacology* 91 (5):380–385. 10.1139/cjpp-2012-0190.

Deshmukh S.A., Barge S.H., Gaikwad D.K. 2014. "Palyno-Morphometric Studies in Some Cassia L. Species from Maharashtra." *Indian Journal of Plant Sciences* 3 (3):71–78.

Deshpande H.A., Bhalsing S.R. 2013. "Recent Advances in the Phytochemistry of Some Medicinally Important Cassia Species: A Review." *IJPMBS* 2 (3): 60–78.

Dhanasekaran, Muniyappan, Savarimuthu Ignacimuthu, Paul Agastian. 2009. "Potential Hepatoprotective Activity of Ononitol Monohydrate Isolated from *Cassia tora* L. on Carbon Tetrachloride Induced Hepatotoxicity in Wistar Rats." *Phytomedicine* 16 (9):891–895.

Dong, Xiaoxv, Yawen Zeng, Yi Liu, Longtai You, Xingbin Yin, Jing Fu, Jian Ni. 2020. "Aloe-Emodin: A Review of Its Pharmacology, Toxicity, and Pharmacokinetics." *Phytotherapy Research*. John Wiley and Sons Ltd. 10.1002/ptr.6532.

El Senousy A.S., Farag M.A., Al-Mahdy D.A., Wessjohann L.A. 2014. "Developmental Changes in Leaf Phenolics Composition from Three Artichoke cvs. (*Cynara scolymus*) as Determined via UHPLC-MS and Chemometrics." *Phytochemistry* 108:67–76.

Farooq M.O., Aziz M.A., Ahmad M.S. 1956. "Seed Oils from *Cassia fistula*, *C. occidentalis*, and *C. tora* (Indian Varieties)." *J Am Oil Chem Soc* 33:21–23.

Fatokun, Amos A., Mercedes Tome, Robert A. Smith, L. Gail Darlington, Trevor W. Stone. 2015. "Protection by the Flavonoids Quercetin and Luteolin against Peroxide-or Menadione-Induced Oxidative Stress in MC3T3-E1 Osteoblast Cells." *Natural Product Research* 29 (12):1127–1132. 10.1080/14786419.2014.980252.

Fei, Junliang, Bin Liang, Chunzhi Jiang, Haifeng Ni, Liming Wang. 2019. "Luteolin Inhibits IL-1β-Induced Inflammation in Rat Chondrocytes and Attenuates Osteoarthritis Progression in a Rat Model." *Biomedicine and Pharmacotherapy* 109 (January):1586–1592. 10.1016/j.biopha.2018.09.161.

Feng, Kai, Zhaoxun Chen, Liu Pengcheng, Shuhong Zhang, Xiaoqing Wang. 2019. "Quercetin Attenuates Oxidative Stress-Induced Apoptosis via SIRT1/AMPK-Mediated Inhibition of ER Stress in Rat Chondrocytes and Prevents the Progression of Osteoarthritis in a Rat Model." *Journal of Cellular Physiology* 234 (10):18192–18205. 10.1002/jcp.28452.

Fricker S.P., Buckley R.G. 1995. "Comparison of Two Colorimetric Assays as Cytotoxicity Endpoints for an In Vitro Screen for Antitumour Agents." *Anticancer Research* 16:3755–3760.

Gardi, C., Bauerova K., Stringa B., Kuncirova V., Slovak L., Ponist S., F. Drafi, et al. 2015. "Quercetin Reduced Inflammation and Increased Antioxidant Defense in Rat Adjuvant Arthritis." *Archives of Biochemistry and Biophysics* 583 (October):150–157. 10.1016/j.abb.2015.08.008.

Gokhale, Jayanti P., Hitendra S. Mahajan, Sanjay S. Surana. 2019. "Quercetin Loaded Nanoemulsion-Based Gel for Rheumatoid Arthritis: In Vivo and in Vitro Studies." *Biomedicine and Pharmacotherapy* 112 (April). 10.1016/j.biopha.2019.108622.

Goto, Tadashi, Keitaro Hagiwara, Nobuaki Shirai, Kaoru Yoshida, Hiromi Hagiwara. 2015. "Apigenin Inhibits Osteoblastogenesis and Osteoclastogenesis and Prevents Bone Loss in Ovariectomized Mice." *Cytotechnology* 67 (2):357–365. 10.1007/s10616-014-9694-3.

Guidline, I.H.T. 2005. "Q2B (R1): Validation of Analytical Procedures: Text and Methodology." Incorporated in November.

Guo, Qiang, Yuxiang Wang, Dan Xu, Johannes Nossent, Nathan J. Pavlos, Jiake Xu. 2018. "Rheumatoid Arthritis: Pathological Mechanisms and Modern Pharmacologic Therapies." *Bone Research*. Sichuan University. 10.1038/s41413-018-0016-9.

Guo, Ava J., Roy C. Choi, Ken Y. Zheng, Vicky P. Chen, Tina T. Dong, Zheng Tao Wang, Günter Vollmer, David T. Lau, Karl W. keung Tsim. 2012. "Kaempferol as a Flavonoid Induces Osteoblastic Differentiation via Estrogen Receptor Signaling." *Chinese Medicine* 7 (April). 10.1186/1749-8546-7-10.

Gupta M., Mazumder U.K., Rath N., Mukhopadhyay D.K. 2000. "Antitumor Activity of Methanolic Extract of *Cassia fistula* L. Seed Against Ehrlich Ascites Carcinoma." *Journal of Ethnopharmacology* 72:151–156.

Gupta R.K. 2010. *Medicinal & Aromatic Plants*. 1st ed. New Delhi:CBS Publishers & Distributors, pp. 116–117.

Hafez Safaa A., Samir M. Osman, Haitham A. Ibrahim, Ahmed A. Seada, Nahla Ayoub. 2019. "Chemical Constituents and Biological Activities of Cassia Genus: Review." *Archives of Pharmaceutical Sciences Ain Shams University* 3 (2):195–227.

Haleagrahara, Nagaraja, Kelly Hodgson, Socorro Miranda-Hernandez, Samuel Hughes, Anupama Bangra Kulur, Natkunam Ketheesan. 2018. "Flavonoid Quercetin–Methotrexate Combination Inhibits Inflammatory Mediators and Matrix Metalloproteinase Expression, Providing Protection to Joints in Collagen-Induced Arthritis." *Inflammopharmacology* 26 (5):1219–1232. 10.1007/s10787-018-0464-2.

Hanley, D.A., Adachi J.D., Bell A., Brown V.. 2012. "Denosumab: Mechanism of Action and Clinical Outcomes." *International Journal of Clinical Practice* 66 (12): 1139–1146. 10.1111/ijcp.12022.

Harshal A., Pawar P. 2011. "Mello Cassia Species Linn: An Overview." *International Journal of Pharmaceutical Sciences and Research* 2 (9):2286–2291.

Hartmann, Kerstin, Mascha Koenen, Sebastian Schauer, Stephanie Wittig-Blaich, Mubashir Ahmad, Ulrike Baschant, Jan P. Tuckermann. 2016. "Molecular Actions of Glucocorticoids in Cartilage and Bone During Health, Disease, and Steroid Therapy." *Physiological Reviews* 96 (2):409–447. 10.1152/physrev.00011.2015.

Hatano, Tsutomu, Seiki Mizuta, Hideyuki Ito, Takashi Yoshida. 1999. "C-Glycosidic Flavonoids from *Cassia occidentalis*." *Phytochemistry* 52 (7):1379–1383.

Hatano T., Uebayashi H., Ito H., et al. 1999. "Phenolic Constituents of *Cassia* Seeds and Antibacterial Effect of Some Naphthalenes and Anthraquinones on Methicillin-Resistent *Staphylococcus aureus*." *Chem Pharm Bull* 47 (8):1121–1127.

Hooker J.D. 1879. *The Flora of British India*. England: L.Reeve and Co., p. 26. Kirtikar, K.R. and Basu, B.D., Indian Medicinal Plants, Vol II, Periodical Experts.

Hou, Yanan, Jiacai Wu, Qin Huang, Lihe Guo. 2009. "Luteolin Inhibits Proliferation and Affects the Function of Stimulated Rat Synovial Fibroblasts." *Cell Biology International* 33 (2):135–147. 10.1016/j.cellbi.2008.10.005.

Hu, Yue, Zhipeng Gui, Yuning Zhou, Lunguo Xia, Kaili Lin, Yuanjin Xu. 2019. "Quercetin Alleviates Rat Osteoarthritis by Inhibiting Inflammation and Apoptosis of Chondrocytes, Modulating Synovial Macrophages Polarization to M2 Macrophages." *Free Radical Biology and Medicine* 145 (December):146–160. 10.1016/j.freeradbiomed.2019.09.024.

Hwang, Jin-Ki, Eun-Mi Noh, Su-Jeong Moon, Jeong-Mi Kim, Kang-Beom Kwon, Byung-Hyun Park, Yong-Ouk You, Bo-Mi Hwang, Hyeong-Jin Kim, Byeong-Soo Kim, Seung-Jin Lee, Jong-Suk Kim, Young-Rae Lee. 2013. "Emodin

Suppresses Inflammatory Responses and Joint Destruction in Collagen-Induced Arthritic Mice." *Rheumatology* 52 (9):1583–1591.

Ilias, Ioannis, Emanouil Zoumakis, Hans Ghayee. 2000. "An Overview of Glucocorticoid Induced Osteoporosis." Endotext. MDText.com, Inc. http://www.ncbi.nlm.nih.gov/pubmed/25905202.

Impellizzeri, Daniela, Emanuela Esposito, Rosanna Di Paola, Akbar Ahmad, Michela Campolo, Angelo Peli, Valeria M. Morittu, Domenico Britti, Salvatore Cuzzocrea. 2013. "Palmitoylethanolamide and Luteolin Ameliorate Development of Arthritis Caused by Injection of Collagen Type II in Mice." *Arthritis Research and Therapy* 15 (6). 10.1186/ar4382.

Irrera, Natasha, Angela D'ascola, Giovanni Pallio, Alessandra Bitto, Emanuela Mazzon, Federica Mannino, Violetta Squadrito, et al. 2019. "β-Caryophyllene Mitigates Collagen Antibody Induced Arthritis (CAIA) in Mice Through a Cross-Talk between CB2 and PPAR-γ Receptors." *Biomolecules* 9 (8). 10.3390/biom9080326.

Jain S.K. 1968. *Medicinal Plants*. New Delhi: National Book Trust, p. 37.

Javadi, Fatemeh, Arman Ahmadzadeh, Shahryar Eghtesadi, Naheed Aryaeian, Mozhdeh Zabihiyeganeh, Abbas Rahimi Foroushani, Shima Jazayeri. 2017. "The Effect of Quercetin on Inflammatory Factors and Clinical Symptoms in Women with Rheumatoid Arthritis: A Double-Blind, Randomized Controlled Trial." *Journal of the American College of Nutrition* 36 (1):9–15. 10.1080/07315724.2016.1140093.

Jiang, Lei, Zenggen Liu, Yulei Cui, Yun Shao, Yanduo Tao, Lijuan Mei. 2019. "Apigenin from Daily Vegetable Celery Can Accelerate Bone Defects Healing." *Journal of Functional Foods* 54 (March):412–421. 10.1016/j.jff.2019.01.043.

Jing, Zheng, Changyuan Wang, Qining Yang, Xuelian Wei, Yue Jin, Qiang Meng, Qi Liu, et al. 2019. "Luteolin Attenuates Glucocorticoid-Induced Osteoporosis by Regulating ERK/Lrp-5/GSK-3β Signaling Pathway in Vivo and in Vitro." *Journal of Cellular Physiology* 234 (4):4472–4490. 10.1002/jcp.27252.

Kadarkarai A.D. 2011. "HPTLC Quantification of Flavonoids, Larvicidal and Smoke Repellent Activities of *Cassia occidentalis* L. (Caesalpiniaceae) Against Malarial Vectore *Anopheles stephensi* Lis (Diptera: Culicidae)." *Journal of Phytology* 3:60–72.

Kamagaté M., Koffi C., Kouamé N.M., Akoubet A., Alain N., Yao R., Die H.M. 2014. "Ethnobotany, Phytochemistry, Pharmacology and Toxicology Profiles of *Cassia siamea* Lam." *J. Phytopharmacol* 3 (1):57–76.

Kapoor V.P., Farooqi M.I.H., Kapoor L.D. 1980. "Chemical Investigations of Seed Mucilages from *Cassia species*." *The Indian Forester* 106 (11):810–812.

Kardinaal A.F.M., Morton M.S., Brüggemann-Rotgans I.E.M., Van Beresteijn E.C.H. 1998. "Phyto-Oestrogen Excretion and Rate of Bone Loss in Postmenopausal Women." *European Journal of Clinical Nutrition* 52 (11):850–855. 10.1038/sj.ejcn.1600659.

Keepers Y.P., Pizao P.E., Peters G.J., van Ark-Otte J., Winograd B., Pinedo H.M. 1991. "Comparison of the Sulforhodamine B Protein and Tetrazolium (MTT) Assays for In Vitro Chemosensitivity Testing." *European Journal of Cancer and Clinical Oncology* 27:897–900.

Khare, C.P. 2007. *Indian Medicinal Plants: An Illustrated Dictionary (Google EBook)*. Springer. http://books.google.com/books?id=gMwLwbUwtfkC&pgis=1.

Kharwadkar N., Mayne B., Lawrence J.E., Khanduja V. 2017. "Bisphosphonates and Atypical Subtrochanteric Fractures of the Femur." *Bone and Joint Research*. British Editorial Society of Bone and Joint Surgery. 10.1302/2046-3758.63.BJR-2016-0125.R1.

Khedgikar, Vikram, Priyanka Kushwaha, Jyoti Gautam, Shewta Sharma, Ashwni Verma, Dharmendra Choudhary, Prabhat R. Mishra, Ritu Trivedi. 2016. "Kaempferol Targets Krt-14 and Induces Cytoskeletal Mineralization in Osteoblasts: A Mechanistic Approach." *Life Sciences* 151 (April):207–217. 10.1016/j.lfs.2016.03.009.

Kim M.K., Chung B.C., Yu V.Y., Nam J.H., Lee H.C., Huh K.B., Lim S.K. 2002. "Relationships of Urinary Phyto-Oestrogen Excretion to BMD in Postmenopausal Women." *Clinical Endocrinology* 56 (3):321–328. 10.1046/j.1365-2265.2002.01470.x.

Kim, Tae Ho, Ji Won Jung, Byung Geun Ha, Jung Min Hong, Eui Kyun Park, Hyun Ju Kim, Shin Yoon Kim. 2011. "The Effects of Luteolin on Osteoclast Differentiation, Function in Vitro and Ovariectomy-Induced Bone Loss." *Journal of Nutritional Biochemistry* 22 (1):8–15. 10.1016/j.jnutbio.2009.11.002.

Kim, Ju Young, Yoon Hee Cheon, Sung Chul Kwak, Jong Min Baek, Kwon Ha Yoon, Myeung Su Lee, Jaemin Oh. 2014. "Emodin Regulates Bone Remodeling by Inhibiting Osteoclastogenesis and Stimulating Osteoblast Formation." *Journal of Bone and Mineral Research* 29 (7):1541–1553. 10.1002/jbmr.2183.

Kim, In Ryoung, Seong Eon Kim, Hyun Su Baek, Bok Joo Kim, Chul Hoon Kim, In Kyo Chung, Bong Soo Park, Sang Hun Shin. 2016. "The Role of Kaempferol-Induced Autophagy on Differentiation and Mineralization of Osteoblastic MC3T3-E1 Cells." *BMC Complementary and Alternative Medicine* 16 (1). 10.1186/s12906-016-1320-9.

Kim, Chang Ju, Sang Hun Shin, Bok Joo Kim, Chul Hoon Kim, Jung Han Kim, Hae Mi Kang, Bong Soo Park, In Ryoung Kim. 2018. "The Effects of Kaempferol-Inhibited Autophagy on Osteoclast Formation." *International Journal of Molecular Sciences* 19 (1). 10.3390/ijms19010125.

Kim, Hae Rim, Bo Mi Kim, Ji Yeon Won, Kyung Ann Lee, Hyun Myung Ko, Young Sun Kang, Sang Heon Lee, Kyoung Woon Kim. 2019. "Quercetin, a Plant Polyphenol, Has Potential for the Prevention of Bone Destruction in Rheumatoid Arthritis." *Journal of Medicinal Food* 22 (2):152–161. 10.1089/jmf.2018.4259.

Kumar, Avinash, Girish K. Gupta, Vikram Khedgikar, Jyoti Gautam, Priyanka Kushwaha, Bendangla Changkija, Geet K. Nagar, et al. 2012. "In Vivo Efficacy Studies of Layer-by-Layer Nano-Matrix Bearing Kaempferol for the Conditions of Osteoporosis: A Study in Ovariectomized Rat Model." *European Journal of Pharmaceutics and Biopharmaceutics* 82 (3):508–517. 10.1016/j.ejpb.2012.08.001.

Kumar, Avinash, Anand K. Singh, Abnish K. Gautam, Deepak Chandra, Divya Singh, Bendangla Changkija, Mahendra Pratap Singh, Ritu Trivedi. 2010. "Identification of Kaempferol-Regulated Proteins in Rat Calvarial Osteoblasts during Mineralization by Proteomics." *Proteomics* 10 (9):1730–1739. 10.1002/pmic.200900666.

Kumar A., Tripathi V., Pushpangadan P. 2007. "Random Amplified Polymorphic DNA as Marker for Genetic Variation and Identification of *Senna surattensis* Burm, f. and *Senna sulfurea* DC. ex Collad." *Curr Sci* 98:1146–1150.

Kumar G.V.P., Naga Subrahmanyam S. 2013. "Phytochemical Analysis, In-Vitro Screening for Antimicrobial and Anthelmintic Activity of Combined Hydroalcoholic Seed Extracts of Four Selected Folklore Indian Medicinal Plants." *Sch Res Libr* 5 (1):168–176.

Kwon, So Mi, Suji Kim, No Joon Song, Seo Hyuk Chang, Yu Jin Hwang, Dong Kwon Yang, Joung Woo Hong, Woo Jin Park, Kye Won Park. 2016. "Antiadipogenic and Proosteogenic Effects of Luteolin, a Major Dietary Flavone, Are Mediated by the Induction of DnaJ (Hsp40) Homolog, Subfamily B, Member 1." *Journal of Nutritional Biochemistry* 30 (April):24–32. 10.1016/j.jnutbio.2015.11.013.

Lai Y.H., Ni Y.N., Kokot S. 2010. "Authentication of Cassia Seeds on the Basis of Two-Wavelength HPLC Fingerprinting with the Use of Chemometrics." *Chinese Chemical Letters* 21:213–216.

Latini, Serena, Felicita Pedata. 2001. "Adenosine in the Central Nervous System: Release Mechanisms and Extracellular Concentrations." *Journal of Neurochemistry*. 10.1046/j.1471-4159.2001.00607.x.

Lavanya B., Maheswaran A., Vimal N., Vignesh K., Uvarani K.Y., Varsha R. 2018. "An Overall View of Cassia Species Phytochemical Constituents and Its Pharmacological Uses." *International Journal of Pharmaceutical Science and Research* 3 (1):47–50.

Lebre, Maria Cristina, Sarah L. Jongbloed, Sander W. Tas, Tom J.M. Smeets, Iain B. McInnes, Paul P. Tak. 2008. "Rheumatoid Arthritis Synovium Contains Two Subsets of CD83 -DC-LAMP- Dendritic Cells with Distinct Cytokine Profiles." *American Journal of Pathology* 172 (4):940–950. 10.2353/ajpath.2008.070703.

Lee C.K., Lee P.H., Kuo Y.H. 2001. "The Chemical Constituents from the Aril of *Cassia fistula* L." *Journal of the Chinese Chemical Society* 48:1053–1058.

Lee H.J., Choi J.S., Jung J.S., Kang S.S. 1998. "Alaternin Glucoside Isomer from *Cassia tora*." *Phytochemistry* 49 (5):1403–1404.

Lee, Su Ui, Hye Kyoung Shin, Yong Ki Min, Seong Hwan Kim. 2008. "Emodin Accelerates Osteoblast Differentiation Through Phosphatidylinositol 3-Kinase Activation and Bone Morphogenetic Protein-2 Gene Expression." *International Immunopharmacology* 8 (5):741–747. 10.1016/j.intimp.2008.01.027.

Lems, Willem F., Kenneth Saag. 2015. "Bisphosphonates and Glucocorticoid-Induced Osteoporosis: Cons." *Endocrine* 49 (3):628–634. 10.1007/s12020-015-0639-1.

Li, Chenghai, Kristifor Sunderic, Steven B. Nicoll, Sihong Wang. 2018. "Downregulation of Heat Shock Protein 70 Impairs Osteogenic and Chondrogenic Differentiation in Human Mesenchymal Stem Cells." *Scientific Reports* 8 (1). 10.1038/s41598-017-18541-1.

Li, Xing, Yanping Han, Qingyou Zhou, Hongyu Jie, Yi He, Jiaochan Han, Juan He, Yong Jiang, Erwei Sun. 2016. "Apigenin, a Potent Suppressor of Dendritic Cell Maturation and Migration, Protects against Collagen-Induced Arthritis." *Journal of Cellular and Molecular Medicine* 20 (1):170–180. 10.1111/jcmm.12717.

Liang, Zhiyuan, Chunfeng Ren. 2018. "Emodin Attenuates Apoptosis and Inflammation Induced by LPS through Up-Regulating LncRNA TUG1 in Murine Chondrogenic ATDC5 Cells." *Biomedicine and Pharmacotherapy* 103 (July):897–902. 10.1016/j.biopha.2018.04.085.

Lin, Chin Chung, Ya Ju Chuang, Chien Chih Yu, Jai Sing Yang, Chi Cheng Lu, Jo Hua Chiang, Jing Pin Lin, Nou Ying Tang, An Cheng Huang, Jing Gung Chung. 2012. "Apigenin Induces Apoptosis Through Mitochondrial Dysfunction in U-2 OS Human Osteosarcoma Cells and Inhibits Osteosarcoma Xenograft Tumor Growth in Vivo." *Journal of Agricultural and Food Chemistry* 60 (45):11395–11402. 10.1021/jf303446x.

Liogier, Henri Alain. 1994. *Descriptive Flora of Puerto Rico and Adjacent Islands.* 1a ed. Vol. III. Rìo Piedras P.R.: Editorial de la Universidad de Puerto Rico.

Liu Z.F., Lang Y.I., Li L.I., Liang Z., Deng Y., Fang R.U.I., Meng Q. 2018. "Effect of Emodin on Chondrocyte Viability in an In Vitro Model of Osteoarthritis." *Experimental and Therapeutic Medicine* 16 (6):5384–5389. 10.3892/etm.2018.6877.

Luo, Jing Sheng, Xinhua Zhao, Yu Yang. 2020. "Effects of Emodin on Inflammatory Bowel Disease-Related Osteoporosis." *Bioscience Reports* 40 (1). 10.1042/BSR20192317.

Lutzky, Viviana, Suad Hannawi, Ranjeny Thomas. 2007. "Cells of the Synovium in Rheumatoid Arthritis. Dendritic Cells." *Arthritis Research and Therapy.* 10.1186/ar2200.

Luximon-Ramma, A., Bahorun, T., Soobrattee, M. A., andAruoma, O. I.2002."Antioxidant Activities of Phenolic, Proanthocyanidin, and Flavonoid Components in Extracts of Cassia Fistula."*Journal of Agricultural and Food Chemistry* 50(18): 5042–5047.

Ma, Jian Da, Jun Jing, Jun Wei Wang, Ying Qian Mo, Qian Hua Li, Jian Zi Lin, Le Feng Chen, Lan Shao, Pierre Miossec, Lie Dai. 2019. "Activation of the Peroxisome Proliferator-Activated Receptor γ Coactivator 1β/NFATc1 Pathway in Circulating Osteoclast Precursors Associated with Bone Destruction in Rheumatoid Arthritis." *Arthritis and Rheumatology* 71 (8):1252–1264. 10.1002/art.40868.

Maitya T.K., Mandal S.C., Mukherjee P.K., Saha K., Dass J., Saha B.P., et al. 1997. "Evaluation of Hepatoprotective Potential of Cassia Species Leaf Extract." *Nat. Prod. Sci* 3:122.

Man G.S., Mologhianu G. 2014. "Osteoarthritis Pathogenesis - A Complex Process That Involves the Entire Joint." *Journal of Medicine and Life*. Carol Davila – University Press.

Manogaran S., Sulochana N. 2004. "Anti-Inflammatory Activity of *Cassia auriculata*." *Ancient Science of Life* 24:65–67.

Mazumder U.K., Gupta M., Rath N. 1998. "CNS Activities of *Cassia fistula* in Mice." *Phytotherapy Research* 12:520–522.

Mehta J.P. 2012. "Separation and Characterization of Anthraquinone Derivatives from *Cassia fistula* Using Chromatographic and Spectral Techniques." *Int J Chem Sci* 10:306–316.

Mirza, Faryal, Ernesto Canalis. 2015. "Secondary Osteoporosis: Pathophysiology and Management." *European Journal of Endocrinology*. BioScientifica Ltd. 10.1530/EJE-15-0118.

Morgan, Elise F., Anthony De Giacomo, Louis C. Gerstenfeld. 2014. "Overview of Skeletal Repair (Fracture Healing and Its Assessment)." *Methods in Molecular Biology* 1130:13–31. 10.1007/978-1-62703-989-5_2.

Nash, Leslie A., Philip J. Sullivan, Sandra J. Peters, Wendy E. Ward. 2015. "Rooibos Flavonoids, Orientin and Luteolin, Stimulate Mineralization in Human Osteoblasts through the Wnt Pathway." *Molecular Nutrition and Food Research* 59 (3):443–453. 10.1002/mnfr.201400592.

Ni Y., Lai Y., Brandes S., Kokot S. 2009. "Multi-Wavelength HPLC Fingerprints from Complex Substances: An Exploratory Chemometrics Study of the Cassia Seed Example." *Analytica Chimica Acta* 647:149–158.

Nirmala A., Eliza I., Rajalakshmi M., Priya E., Daisy P. 2008. "Effect of Hexane Extract of *Cassia fistula* Barks on Blood Glucose and Lipid Profile in Streptozotocin Diabetic Rats." *International Journal of Pharmacology* 4:292–296.

Nsonde Ntandou G.F., Banzouzi J.T., Mbatchi B., Elion-Itou R.D.G., Etou-Ossibi A.W., Ramos S., Benoit-Vical F., Abena A.A., Ouamba J.M. 2010. "Analgesic and Anti-Inflammatory Effects of *Cassia siamea* Lam. Stem Bark Extracts." *Journal of Ethnopharmacology* 127:108–111.

Odason, E.E., J.A. Kolawole. 2007. "Of the Aqueous Extract of the Root of *Cassia siamea* Lam. (Ceasalpiniaceae)." *Nigerian Journal of Pharmaceutical Research* 6 (1):66–69.

O'Sullivan, Saoirse Elizabeth. 2016. "An Update on PPAR Activation by Cannabinoids." *British Journal of Pharmacology*. John Wiley and Sons Inc. 10.1111/bph.13497.

Padumanonda, Tanit, Wandee Gritsanapan. 2006. "Barakol Contents in Fresh and Cooked *Senna siamea* Leaves." *Southeast Asian Journal of Tropical Medicine and Public Health* 37 (2):388.

Pal, Subhashis, Padam Kumar, Eppalapally Ramakrishna, Sudhir Kumar, Konica Porwal, Brijesh Kumar, Kamal R. Arya, Rakesh Maurya, Naibedya Chattopadhyay. 2019. "Extract and Fraction of Cassia Occidentalis L. (a Synonym of Senna Occidentalis) Have Osteogenic Effect and Prevent Glucocorticoid-Induced Osteopenia." *Journal of Ethnopharmacology* 235 (May):8–18. 10.1016/j.jep.2019.01.029.

Pal, Subhashis, Naresh Mittapelly, Athar Husain, Sapana Kushwaha, Sourav Chattopadhyay, Padam Kumar, Eppalapally Ramakrishna, et al. 2020. "A Butanolic Fraction from the Standardized Stem Extract of Cassia Occidentalis L Delivered by a Self-Emulsifying Drug Delivery System Protects Rats from Glucocorticoid-Induced Osteopenia and Muscle Atrophy." *Scientific Reports* 10 (1):1–14. 10.1038/s41598-019-56853-6.

Pan, Fang, Lihua Zhu, Haozhe Lv, Chunpeng Pei. 2016. "Quercetin Promotes the Apoptosis of Fibroblast-Like Synoviocytes in Rheumatoid Arthritis by Upregulating LncRNA MALAT1." *International Journal of Molecular Medicine* 38 (5):1507–1514. 10.3892/ijmm.2016.2755.

Pandey R., Chandra P., Arya K.R., Kumar B. 2014. "Development and Validation of an Ultra High Performance Liquid Chromatography Electrospray Ionization Tandem Mass Spectrometry Method for the Simultaneous Determination of Selected Flavonoids in *Ginkgo biloba*." *Journal of Separation Science* 37:3610–3618.

Pandit, Ashlesha P., Sachin B. Omase, Vaishali M. Mute. 2020. "A Chitosan Film Containing Quercetin-Loaded Transfersomes for Treatment of Secondary Osteoporosis." *Drug Delivery and Translational Research*. 10.1007/s13346-020-00708-5.

Pang, Jian L., Dennis A. Ricupero, Su Huang, Nigar Fatma, Dhirendra P. Singh, Jose R. Romero, Naibedya Chattopadhyay. 2006. "Differential Activity of Kaempferol and Quercetin in Attenuating Tumor Necrosis Factor Receptor Family Signaling in Bone Cells." *Biochemical Pharmacology* 71 (6):818–826. 10.1016/j.bcp.2005.12.023.

Paolino, Sabrina, Maurizio Cutolo, Carmen Pizzorni. 2017. "Glucocorticoid Management in Rheumatoid Arthritis: Morning or Night Low Dose?" *Reumatologia*. Termedia Publishing House Ltd. 10.5114/reum.2017.69779.

Pari L., Latha M. 2002. "Effect of *Cassia auriculata* Flowers on Blood Sugar Levels, Serum and Tissue Lipids in Streptozotocin Diabetic Rats." *Singapore Med J* 43:617–621.

Park, Jeong A., Sang Keun Ha, Tong Ho Kang, Myung Sook Oh, Min Hyoung Cho, Soo Yeol Lee, Ji Ho Park, Sun Yeou Kim. 2008. "Protective Effect of Apigenin on Ovariectomy-Induced Bone Loss in Rats." *Life Sciences* 82 (25–26):1217–1223. 10.1016/j.lfs.2008.03.021.

Park K.H., Park J.D., Hyun K.H., Nakayama M., Yokota T. 1994. "Brassinosteroids and Monoglycerides in Immature Seeds of *Cassia species* as the Active Principles in the Rice Lamina Inclination Bioassay." *Biosci Biotechn Biochem* 58 (7):1343–1344.

Parveen, M., M. Kamil, M. Ilyas. 1995. "A New Isoflavone C-Glycoside from *Cassia siamea*." *Fitoterapia* 66 (5):439–441.

Patel, Neeraj K., Pulipaka Sravani, Shashi P. Dubey, Kamlesh K. Bhutani. 2014. "Pro-Inflammatory Cytokines and Nitric Oxide Inhibitory Constituents from *Cassia occidentalis* Roots." *Nat Prod Commun* 9 (5):661–664. https://pubmed.ncbi.nlm.nih.gov/25026715/?from_single_result=25026715&expanded_search_query=25026715.

Patidar P., Dubey D., Dashora K. 2012. "Quantification Of (-) Epicatechin by HPTLC Method In Cassia Fistula Crude Drug, Lab Extract and Commercial Extract." *Asian Journal of Pharmaceutical and Clinical Research* 5:140–143.

Prakash D., Suri S., Upadhyay G., Singh B.N. 2007. "Total Phenol, Antioxidant and Free Radical Scavenging Activities of Some Medicinal Plants." *International Journal of Food Sciences and Nutrition* 58:18–28.

Quan, He, Xiaopeng Dai, Meiyan Liu, Chuanjun Wu, Dan Wang. 2019. "Luteolin Supports Osteogenic Differentiation of Human Periodontal Ligament Cells." *BMC Oral Health* 19 (1). 10.1186/s12903-019-0926-y.

Rius-Pérez, S., Torres-Cuevas I., Millán I., Ortega Á.L., Pérez S. 2020. "PGC-1 α, Inflammation, and Oxidative Stress: An Integrative View in Metabolism." *Oxidative Medicine and Cellular Longevity* 2020. 10.1155/2020/1452696.

Saccol, Renata da Silva Pereira, Karine Lanes da Silveira, Stephen Adeniyi Adefegha, Alessandra Guedes Manzoni, Leonardo Lanes da Silveira, Ana Paula Visintainer Coelho, Livia Gelain Castilhos, et al. 2019. "Effect of Quercetin on E-NTPDase/E-ADA Activities and Cytokine Secretion of Complete Freund Adjuvant–Induced Arthritic Rats." *Cell Biochemistry and Function* 37 (7):474–485. 10.1002/cbf.3413.

Sakulpanich A., Chewchinda S., Sithisarn P., Gritsanapan W. 2012. "Standardization and Toxicity Evaluation of *Cassia fistula* Pod Pulp Extract for Alternative Source of Herbal Laxative Drug." *Pharmacognosy Journal* 4:6–12.

Sartorelli P., Andrade S.P. 2007. "Melhem MrSC, Prado FO, Tempone AG: Isolation of Antileishmanial Sterol from the Fruits of *Cassia fistula* Using Bioguided Fractionation." *Phytotherapy Research* 21:644–647.

Shailajan S., Yeragi M., Tiwari B. 2013. "Estimation of Rhein from *Cassia fistula* Linn. Using Validated HPTLC Method." *International Journal of Green Pharmacy* 7:62.

Shan, Jinghua, Lixia Chen, Keliang Lu. 2017. "Protective Effects of Trans-Caryophyllene on Maintaining Osteoblast Function." *IUBMB Life* 69 (1):22–29. 10.1002/iub.1584.

Sharma A., Ahmad S., Hari Kumar S.L., et al. 2014. "Pharmacognosy, Phytochemistry & Pharmacology of Cassia Javanica Linn.: A Review." *International Journal of Pharma Research & Review* 3(4):101–105.

Sheeba M., Emmanuel S., Revathi K., Ignacimuthu S. 2009. "Wound Healing Activity of *Cassia occidentalis* L. in Albino Wistar Rats." *International Journal of Integrative Biology* 8 (1):1–6.

Shi, Fengchao, Dun Zhou, Zhongqiu Ji, Zhaofeng Xu, Huilin Yang. 2015. "Anti-Arthritic Activity of Luteolin in Freund's Complete Adjuvant-Induced Arthritis in Rats by Suppressing P2X4 Pathway." *Chemico-Biological Interactions* 226 (January):82–87. 10.1016/j.cbi.2014.10.031.

Siddhuraju P., Mohan P.S., Becker K. 2002. "Studies on the Antioxidant Activity of Indian Laburnum (*Cassia fistula* L.): A Preliminary Assessment of Crude Extracts from Stem Bark, Leaves, Flowers and Fruit Pulp." *Food Chemistry* 79:61–67.

Siddiqui, Jawed A., Gaurav Swarnkar, Kunal Sharan, Bandana Chakravarti, Gunjan Sharma, Preeti Rawat, Manmeet Kumar, et al. 2010. "8,8″-Biapigeninyl Stimulates Osteoblast Functions and Inhibits Osteoclast and Adipocyte Functions: Osteoprotective Action of 8,8″-Biapigeninyl in Ovariectomized Mice." *Molecular and Cellular Endocrinology* 323 (2):256–267. 10.1016/j.mce.2010.03.024.

Silva, Barbara C., John P. Bilezikian. 2011. "New Approaches to the Treatment of Osteoporosis." *Annual Review of Medicine* 62 (1):307–322. 10.1146/annurev-med-061709-145401.

Silva, Mirtes G.B., Ticiana P. Aragão, Carlos F.B. Vasconcelos, Pablo A. Ferreira, Bruno A. Andrade, Igor M.A. Costa, João H. Costa-Silva, Almir G. Wanderley, Simone S.L. Lafayette. 2011. "Acute and Subacute Toxicity of *Cassia occidentalis* L. Stem and Leaf in Wistar Rats." *Journal of Ethnopharmacology* 136 (2):341–346. 10.1016/j.jep.2011.04.070.

Singh, Vibha. 2017. "Medicinal Plants and Bone Healing." *National Journal of Maxillofacial Surgery* 8 (1):4. 10.4103/0975-5950.208972.

Singh V.K., Khan A.M. 1990. *Medicinal Plants and Folklores - A Strategy towards Conquest of Human Ailments.* Vol. 9. Today & Tomorrow Printers & Publishers, p. 67.

Singh P., Karnwal P.P. 2006. "Antifungal Activity of *Cassia fistula* Leaf Extract Against *Candida albicans.*" *Indian Journal of Microbiology* 46:169.

Singh S., Singh S.K., Yadav A. 2013. "Review on CASSIA Species: Pharmacological, Traditional and Medicinal Aspects in Various Countries." *American Journal of Phytomedicine and Clinical Therapeutics* 1 (3):291–312.

Singh, Harpreet, Piyush Chahal, Amrita Mishra, Arun Kumar Mishra. 2019. "An Up-to-Date Review on Chemistry and Biological Activities of *Senna occidentalis* (L.) Link Family: Leguminosae." *Oriental Pharmacy and Experimental Medicine*: 1–16. 10.1007/s13596-019-00391-z.

Skehan P., Storeng R., Scudiero D., Monks A., McMahon J., Vistica D., Warren J.T., Bokesch H., Kenney S., Boyd M.R. 1990. "New Colorimetric Cytotoxicity Assay for Anticancer-Drug Screening." *Journal of the National Cancer Institute* 82:1107–1112.

Sozen, Tumay, Lale Ozisik, Nursel Calik Basaran. 2017. "An Overview and Management of Osteoporosis." *European Journal of Rheumatology* 4 (1):46–56. 10.5152/eurjrheum.2016.048.

Stevens, Warren Douglas. 2001. *Flora de Nicaragua.* St. Louis, MO: Missouri Botanical Garden Press.

Suh, Kwang Sik, Eun Mi Choi, Mikwang Kwon, Suk Chon, Seungjoon Oh, Jeong Taek Woo, Sung Woon Kim, Jin Woo Kim, Young Seol Kim. 2009. "Kaempferol Attenuates 2-Deoxy-D-Ribose-Induced Oxidative Cell Damage in MC3T3-E1 Osteoblastic Cells." *Biological and Pharmaceutical Bulletin* 32 (4):746–749. 10.1248/bpb.32.746.

Tabacco, Gaia, John P. Bilezikian. 2019. "Osteoanabolic and Dual Action Drugs." *British Journal of Clinical Pharmacology.* Blackwell Publishing Ltd. 10.1111/bcp.13766.

Tang, Xiaolu, Xiaoyan Zhu, Shujuan Liu, Richard C. Nicholson, Xin Ni. 2008. "Phytoestrogens Induce Differential Estrogen Receptor β-Mediated Responses in Transfected MG-63 Cells." *Endocrine* 34 (1–3):29–35. 10.1007/s12020-008-9099-1.

Taylor, Adam D., Kenneth G. Saag. 2019. "Anabolics in the Management of Glucocorticoid-Induced Osteoporosis: An Evidence-Based Review of Long-Term Safety, Efficacy and Place in Therapy." *Core Evidence* 14 (August):41–50. 10.2147/ce.s172820.

Teitelbaum S.L. 2016. "Therapeutic Implications of Suppressing Osteoclast Formation Versus Function." *Rheumatology (Oxford)* 55 (suppl 2):61–63. https://pubmed.ncbi.nlm.nih.gov/27856662/?from_term=Therapeutic+implications+of+suppressing+osteoclast+formation+versus+function&from_pos=1.

Tona L., Mesia K., Ngimbi N.P., Chrimwami B., Okond' A., Cimanga K., De Bruyne T. et al. 2001. "In-Vivo Antimalarial Activity of *Cassia occidentalis*, *Morinda morindoides* and *Phyllanthus niruri.*" *Annals of Tropical Medicine & Parasitology* 95 (1):47–57.

Tona L., Ngimbi N.P., Tsakala M., Mesia K., Cimanga K., Apers S., De Bruyne T., Pieters L., Totte J., Vlietinck A.J. 1999. "Antimalarial Activity of 20 Crude

Extracts from Nine African Medicinal Plants Used in Kinshasa, Congo." *Journal of Ethnopharmacology* 68 (1–3):193–203.

Tsuji, Mitsuyoshi, Hironori Yamamoto, Tadatoshi Sato, Yoko Mizuha, Yoshichika Kawai, Yutaka Taketani, Shigeaki Kato, Junji Terao, Takahiro Inakuma, Eiji Takeda. 2009. "Dietary Quercetin Inhibits Bone Loss without Effect on the Uterus in Ovariectomized Mice." *Journal of Bone and Mineral Metabolism* 27 (6):673–681. 10.1007/s00774-009-0088-0.

Tzakou O., Loukis A., Said A. 2007. "Essential Oil from the Flowers and Leaves of Cassia fistula L." *Journal of Essential Oil Research* 19:360–361.

Van Beek E.R., Löwik C.W.G.M., Papapoulos S.E. 2002. "Bisphosphonates Suppress Bone Resorption by a Direct Effect on Early Osteoclast Precursors without Affecting the Osteoclastogenic Capacity of Osteogenic Cells: The Role of Protein Geranylgeranylation in the Action of Nitrogen Containing Bisphosphonates on Osteoclast Precursors." *Bone* 30 (1):64–70. 10.1016/S8756-3282(01)00655-X.

Veerachari U., Bopaiah A.K. 2012. "Phytochemical Investigation of the Ethanol, Methanol and Ethyl Acetate Leaf Extracts of Six Cassia Species." *International Journal of Pharma and Bio Sciences* 2 (2):260–270.

Wadekar J., Sawant R., Honde B. 2011. "Anthelmintic and Antibacterial Potential of *Cassia auriculata* Roots." *Int J Pharm Front Res* 1:93–98.

Wagner, Hildebert, Samia Mohammed El-Sayyad, Otto Seligmann, V. Mohan Chari. 1978. "Chemical Constituents of *Cassia siamea* Lam., I." *Planta Medica* 33 (03):258–261.

Wattel, Alice, Said Kamel, Romuald Mentaverri, Florence Lorget, Christophe Prouillet, Jean Pierre Petit, Patrice Fardelonne, Michel Brazier. 2003. "Potent Inhibitory Effect of Naturally Occurring Flavonoids Quercetin and Kaempferol on In Vitro Osteoclastic Bone Resorption." *Biochemical Pharmacology* 65 (1):35–42. 10.1016/S0006-2952(02)01445-4.

Wei S-y., Yao W-x., Ji W-y., Wei J-q., Peng S-q. 2013. "Qualitative and Quantitative Analysis of Anthraquinones in Rhubarbs by High Performance Liquid Chromatography with Diode Array Detector and Mass Spectrometry." *Food Chemistry* 141:1710–1715.

Wei, Bing, Yan Zhang, Lixia Tang, Yikui Ji, Cheng Yan, Xiaoke Zhang. 2019. "Protective Effects of Quercetin Against Inflammation and Oxidative Stress in a Rabbit Model of Knee Osteoarthritis." *Drug Development Research* 80 (3):360–367. 10.1002/ddr.21510.

Wenink, M.H., W. Han, R.E.M. Toes, T.R.D.J. Radstake. 2008. "Dendritic Cells and Their Potential Implication in Pathology and Treatment of Rheumatoid Arthritis." *In Dendritic Cells*, 81–98. Springer Berlin Heidelberg. 10.1007/978-3-540-71029-5_4.

Wong, Sok Kuan, Kok Yong Chin, Soelaiman Ima-Nirwana. 2019. "The Osteoprotective Effects of Kaempferol: The Evidence from In Vivo and In Vitro Studies." *Drug Design, Development and Therapy*. Dove Medical Press Ltd. 10.2147/DDDT.S227738.

Xia Y., Wei G., Si D., Liu C. 2011. "Quantitation of Ursolic Acid in Human Plasma by Ultra Performance Liquid Chromatography Tandem Mass Spectrometry and Its Pharmacokinetic Study." *Journal of Chromatography B* 879:219–224.

Yadav J.P., Arya V., Yadav S., Panghal M., Kumar S., Dhankhar S. 2010. "*Cassia occidentalis* L.: A Review on Its Ethnobotany, Phytochemical and Pharmacological Profile." *Fitoterapia* 81 (4):223–230.

Yadav R., Jain G.C. 2009. "Antifertility Effect and Hormonal Profile of Petroleum Ether Extract of Seeds of *Cassia fistula* in Female Rats." *International Journal of Pharmacology and Technology Research* 1:438–444.

Yamaguchi, Masayoshi, Robert M. Levy. 2016. "β-Caryophyllene Promotes Osteoblastic Mineralization, and Suppresses Osteoclastogenesis and Adipogenesis in Mouse Bone Marrow Cultures in Vitro." *Experimental and Therapeutic Medicine* 12 (6):3602–3606. 10.3892/etm.2016.3818.

Yang, Feng, Pu wei Yuan, Yang Quan Hao, Zheng Mao Lu. 2014. "Emodin Enhances Osteogenesis and Inhibits Adipogenesis." *BMC Complementary and Alternative Medicine* 14 (February). 10.1186/1472-6882-14-74.

Yang, Shun Chin, Po Jen Chen, Shih Hsin Chang, Yu Ting Weng, Fang Rong Chang, Kuang Yi Chang, Chun Yu Chen, Ting I. Kao, Tsong Long Hwang. 2018. "Luteolin Attenuates Neutrophilic Oxidative Stress and Inflammatory Arthritis by Inhibiting Raf1 Activity." *Biochemical Pharmacology* 154 (August):384–396. 10.1016/j.bcp.2018.06.003.

Yang, Yan, Xu Zhang, Min Xu, Xin Wu, Feipeng Zhao, Chengzhi Zhao. 2018. "Quercetin Attenuates Collagen-Induced Arthritis by Restoration of Th17/Treg Balance and Activation of Heme Oxygenase 1-Mediated Anti-Inflammatory Effect." *International Immunopharmacology* 54 (January):153–162. 10.1016/j.intimp.2017.11.013.

Yegutkin, Gennady G. 2008. "Nucleotide- and Nucleoside-Converting Ectoenzymes: Important Modulators of Purinergic Signalling Cascade." *Biochimica et Biophysica Acta - Molecular Cell Research*. 10.1016/j.bbamcr.2008.01.024.

Yu Q., Xiang J., Tang W., Liang M., Qin Y., Nan F. 2009. "Simultaneous Determination of the 10 Major Components of Da-Cheng-Qi Decoction in Dog Plasma by Liquid Chromatography Tandem Mass Spectrometry." *Journal of Chromatography B* 877:2025–2031.

Zhang, Xue, Chenhui Zhou, Xuan Zha, Zhoumei Xu, Li Li, Yuyu Liu, Liangliang Xu, Liao Cui, Daohua Xu, Baohua Zhu. 2015. "Apigenin Promotes Osteogenic Differentiation of Human Mesenchymal Stem Cells through JNK and P38 MAPK Pathways." *Molecular and Cellular Biochemistry* 407 (1–2):41–50. 10.1007/s11010-015-2452-9.

Zhao W., Zeng X., Zhang T., Wang L., Yang G., Chen Y.-K., Hu Q., Miao M. 2013. "Flavonoids from the Bark and Stems of *Cassia fistula* and Their Anti-tobacco Mosaic Virus Activities." *Phytochemistry Letters* 6:179–182.

Zhou, Manru, Jin Li, Jingkai Wu, Yajun Yang, Xiaobing Zeng, Xiaohua Lv, Liao Cui, Weimin Yao, Yuyu Liu. 2017. "Preventive Effects of Polygonum Multiflorum on Glucocorticoid-Induced Osteoporosis in Rats." *Experimental and Therapeutic Medicine* 14 (3):2445–2460. 10.3892/etm.2017.4802.

Index

A

B

D

E

F

G

H

I

J

K

L

M

N

Q

R

S